FINDING & CORRECTING ERRORS

Whole Number Computation
Addition and Subtraction

A Skills Trajectory Approach

FINDING & CORRECTING ERRORS

Whole Number Computation
Addition and Subtraction

E.T. Walker, Ph.D.

Math Explorations

Printed in the United States of America.

ISBN (paperback): 978-1-7363939-0-1
ISBN (PDF version): 978-1-7363939-1-8

Contents

Preface

Like many educators, I am fascinated with how we learn and how that learning culminates in a skill or final conceptual understanding. When we arrive at a learning destination and the skill has been internalized, the brain efficiently consolidates and forgets the building blocks or steps along the way. Reflecting on teaching my own children to tie their shoes or drive a car, I remember that moment of celebration when the learning became automatic. The shoe was tied without repeating each step out loud. The car ride was pleasant, safe, and maybe accompanied by a conversation!

Within learning, though, there is the possibility for missed steps, unheard directions, or misunderstandings. Unaware, we all carry these small gaps in our knowledge. Their illumination is sometimes accidental; sometimes guided by a more knowledgeable other. Finding a buried misconception in my own learning always elicits a mixture of delight and satisfaction as the missing piece falls into place.

Our students are faced with learning enormous amounts of information. The potential to be distracted, to misinterpret, or be absent the day something is taught is high. Resulting gaps may be crucial to later success. With the help of teachers and students, I have explored a systematic way to look for those small, crucial gaps in computational knowledge, which keep students from being successful. It is my hope that identifying these gaps will empower teachers and parents to remediate efficiently and filling in the missing pieces will empower students to see themselves as math-able and successful.

This method is not about memorizing math facts nor about which computational method is best. Using a skills trajectory is about identifying and remediating gaps or misconceptions. Within a skills trajectory, teachers or parents have the freedom to teach each step however works best for their students. Teachers and parents may teach as much conceptual understanding of each level as desired or needed.

Finding and Correcting Errors in Computation using Skills Trajectories

Learning anything is an amazing feat and the brain handles most learning with incredible efficiency. Once automaticity of a skill is acquired, the step-by-step learning that led us there is discarded. To teach another, end goals must be *deconstructed* into basic steps, which can be reconstructed or taught in a logical sequence. These basic steps that lead to an end skill are called a skills trajectory.

Accurately adding, subtracting, multiplying, and dividing are foundational skills taught at the elementary school level. These skills, while supplanted by calculators in the middle school, provide the foundations for reasoning mathematically, looking for patterns, performing operations on fractions, and estimating.

Addressing gaps in understanding of mathematical computations often takes the form of returning to a recent lesson or the beginning of a grade level body of knowledge. This assumes that the content of prior grades has been mastered and errors lie in understanding current or recent instruction.

For many students, conceptual misunderstandings or gaps from earlier grades create weak and possibly unbridgeable links to current grade level knowledge. The grade level gaps in understanding is where remediation should begin. Initial trials using this method found 5th graders, who struggled with operations on decimals, were lacking procedural knowledge of regrouping from 2nd grade content. Direct instruction, tailored to these needs, immediately improved students' mathematical ability at their current grade level.

Finding gaps in understanding, the small and sometimes personal misconceptions or missing concepts, is difficult. Using a preassessment of steps in a skills trajectory is an efficient method for determining where student knowledge begins to fall apart. Learning opportunities and practice at each step help students progress along the trajectory toward an end goal.

A skills trajectory provides:

- an end goal: in this case being able to add and subtract whole numbers,
- a pathway or trajectory of the smallest components, steps, or levels of that goal
- a diagnostic tool for determining a student's point of entry along this pathway and to see if learning was sticky when remediation is complete,
- activities or lessons which move students from one step to the next along the trajectory.

I am not a proponent of computation as the end all of math education. I am a proponent of student and teacher success. The use of skills trajectories in remediation and instruction provides success for all.

Using this Resource

A. Each computational area has its own section which contains:

1. *A skills trajectory or set of steps.* The skills trajectory is comprised of numbered steps or levels. Grade level expectations are imbedded but are for guidance, only. Order of steps matters but is not absolute.
2. *A pre/post assessment.* Each item on the assessment corresponds with a numbered level on the trajectory. For example: Item 1 on the Whole Number Addition Pre/Post Assessment corresponds to Level 1 of the Whole Number Addition Trajectory.
3. A tracker, which can be recreated in a spreadsheet or word document to organize the pre/post data and document growth.
4. A list of common errors within the trajectory.
5. Two practice sheets for each level. Mixed practice sheets for every two levels.
6. Answer keys for all pre/post assessments and practice sheets.

B. How to proceed

Pre-assessing and grouping

1. Pre-assess using the Pre/Post Assessment. It is suggested that you begin with the addition trajectory. All students should begin the assessment at Level 1. All students should stop when they feel they are becoming frustrated or can no longer complete problems.
2. Score and record each item on the Pre/Post Assessment using a 1 for each item that is completely correct and a 0 for those with any kind of error.
3. Group students by where they fall on the trajectory. Do you have a group that seems to be stuck at or around Level 7 and another group that is stuck at or around Level 11?
4. Find time to work in small groups. There are many ways to do this, so be creative. Compare this work to guided reading groups. A ten minute, mini-lesson followed by 10 minutes of practice is often sufficient. Begin with the level where the scores of 0 begin.
5. Use your own judgement, but it is advised to not skip levels.
6. Post-assess when all levels up to grade level expectations in the trajectory have been mastered. You may wish to use levels beyond the current grade as enrichment.
7. Return to and review any levels where students continue to have misconceptions.

Instruction and practice

1. Keep the instructional group moving through the trajectory together. Otherwise, multiple students will be working on different levels, which becomes complicated.
2. The teacher should have a firm understanding of the skill and common errors for each level. Errors tend to fall into categories. Once pre-assessments have been scored and trajectory levels have been identified for a group starting point, sort student work on the pre-assessment by error type. Make a list of these errors to address in your guided mini-lesson.
3. The teacher creates and engages students in a guided mini-lesson for the current level.
4. Students complete the first practice problem ONLY and check in with the instructor.
5. The teacher checks progress on just the first problem and may choose to highlight two or three solutions in a discussion or number talk if students are still unsure of the skill. Misunderstandings are addressed immediately.
6. When the teacher is confident that students have some understanding of the skill, students complete two more problems and check in with the group and teacher. Misunderstandings are again addressed immediately.
7. Students are directed to complete the rest of the practice sheet when understanding seems solid and errors are rectified. Student progress should be observed by a teacher as students progress through the practice set.
8. The teacher may choose to have students complete a motivational Level Tracker of their own, starring or checking off levels they have mastered. In this era of online games, leveling up is highly desired by students.
9. *Handing out practice sheets without this procedure is not recommended.* The repetition of an error ingrains a misconception in the student's brain. It is imperative to catch and remediate errors immediately through small group instruction and monitoring before they are repeated on multiple problems.
10. The second practice sheet may be used at the discretion of the teacher or parent. Ideally, the second practice sheet should not be completed the same day. Similarly, the mixed practice sheets may be used at the discretion of the instructor.

Addition Trajectory

Addition Skills Trajectory
Multi-Digit Addition for Whole Numbers

(Levels on the trajectory match the numbers on the Multi-Digit Addition Pre- and Post-Assessment)

<u>Kindergarten begins here</u>
Assumptions

- ✔ Understanding of the concept of addition (putting together)
- ✔ Showing addition as combining groups using manipulatives or models

Level 1 – Counting on within 100. I say 14, you say the next number. I say 21 you say the next 2 (or 3) numbers both within tens and across tens.

Level 2 – Adding single digits within 10 both horizontally and vertically with or without manipulatives (fingers are fine)

<u>1st grade begins here</u>

Level 3 – Two-digit + one-digit, no regrouping
Level 4 – Two-digit + two-digit, no regrouping
Level 5 – Two-digit + one-digit, regrouping one's place
Level 6 – Two-digit + two-digit, regrouping one's place

<u>2nd Grade begins here</u>

Level 7 – Two-digit + two-digit, regrouping ten's place, only
Level 8 – Two-digit + two-digit, regrouping BOTH one's and ten's place
Level 9 – Three-digit + one-digit with and without regrouping
Level 10 – Three-digit + two-digit, with and without regrouping one place value
Level 11 – Three-digit + three-digit, no regrouping
Level 12 – Three-digit + three-digit, regrouping one's or ten's place

<u>3rd Grade begins here</u>

Level 13 – Three-digit + three-digit, regrouping two place values
Level 14 – Three- or four-digit + three-digit, regrouping into the thousand's place

Pre and Post Assessment for Adding Multi-Digit Numbers

Name: ______________________________ Date: ____________________

1. Count forward: 45, ___, ___, ___, 49	**2.** Add: 4 + 3 = _____	**3.** Add: 12 + 3
4. Add: 14 + 21	**5.** Add: 27 + 5	**6.** Add: 46 + 27
7. Add: 62 + 81	**8.** Add: 76 + 87	

Pre and Post Assessment for Adding Multi-Digit Numbers

Name: ______________________________ Date: ____________________

9. Add: $456 + 3$ $456 + 9$	**10.** Add: $451 + 14$ $451 + 84$	**11.** Add: $235 + 352$
12. Add: $238 + 314$	**13.** Add: $278 + 354$	**14.** Add: $512 + 876$

Pre and Post Assessment for Adding Multi-Digit Numbers
ANSWER KEY

Name: ______________________ Date: ______________________

1.	2.	3.
Count forward: 45, 46, 47, 48, 49	Add: 4 + 3 = 7	Add: 12 + 3 15
4. Add: 14 + 21 35	**5.** Add: 27 + 5 32	**6.** Add: 46 + 27 73
7. Add: 62 + 81 143	**8.** Add: 76 + 87 163	

Pre and Post Assessment for Adding Multi-Digit Numbers
ANSWER KEY

Name: ______________________ Date: ______________________

9.	**10.**	**11.**
Add: 456 + 3 459 456 + 9 465	Add: 451 +14 465 451 +84 535	Add: 235 +352 587
12.	**13.**	**14.**
Add: 238 +314 552	Add: 278 +354 632	Add: 512 +876 1,388

Common Errors
Addition – Whole Numbers

Analysis of student computation errors will often reveal the root of students' misconceptions. When many students in a class are making the same error, the teacher may want to examine the method being used or the explanation given to students for that step. As teachers, what we say is perfectly clear to us, but students may interpret what is said differently.

Misconceptions about the concept of addition

- No understanding that adding means putting parts or numbers together
- Difficulty moving between horizontal and vertical format

Misconceptions about modeling

- Difficulty counting forward from a given number
- Difficulty counting forward across a multiple of 10
- Catching and counting incorrectly, or using a number line incorrectly resulting in the sum being one too great or one less than the correct sum
- Difficulty aligning digits in a place value column

Misconceptions about regrouping

- No regrouping, writing all the digits on the answer line
- "Losing" the regrouped digit, writing only the one's digit in the place value column
- Placing the regrouped digit in the incorrect place value column
- Regrouping correctly but not adding the regrouped digit onto the sum
- Regrouping in one place value but not being able to regroup in two place values
- Unable to regroup if there are no digits in the column to the left

Addition Skills Trajectory Tracker: Whole Numbers

Student	Level 1 Counting on within 100	Level 2 Adding single digits within 10 vertically and horizontally	Level 3 Two digit + one digit, no regrouping	Level 4 Two digit + two digit, no regrouping	Level 5 Two digit + one digit, regrouping one's place	Level 6 Two digit + two digit, regrouping one's place	Level 7 Two digit + two digit, regrouping ten's place, only	Level 8 Two digit + two digit, regrouping BOTH one's and ten's places	Level 9 Three digit + one digit with and without regrouping	Level 10 Three digit + two digit, with and without regrouping one place value	Level 11 Three digit + three digit, no regrouping	Level 12 Three digit + three digit, regrouping one's or ten's place	Level 13 Three digit + three digit, regrouping two place values	Level 14 Three digit + three digit, regrouping into the thousand's place

Addition Trajectory Leveled Practice Sheets

Level 1a Practice Adding Multi-Digit Numbers (Counting on within 100)

Name: ______________________ Date: ______________________

A. Count forward: 4, ____, ____, ____, 8	**B.** Count forward: 12, ____, ____, ____,16	**C.** Count forward: 21, ____, ____, ____, 25
D. Count forward: 45, ____, ____, ____, 49	**E.** Count forward: 68, ____, ____, ____, 72	**F.** Count forward: 99, ____, ____, ____, 103

Level 1b Practice Adding Multi-Digit Numbers (Counting on within 100)

Name: ______________________ Date: ______________________

A. Count forward: 2, ____, ____, ____, 6	**B.** Count forward: 10, ____, ____, ____, 14	**C.** Count forward: 23, ____, ____, ____, 27
D. Count forward: 41, ____, ____, ____, 45	**E.** Count forward: 67, ____, ____, ____, 71	**F.** Count forward: 98, ____, ____, ____, 102

Level 2a Practice Adding Multi-Digit Numbers (Adding single digits within 10, horizontally and vertically)

Name: ______________________________ Date: ______________________

A.	**B.**	**C.**
Add: 3 + 1 = ____	Add: 4 + 3 = ____	Add: 2 + 7 ____
D.	**E.**	**F.**
Add: 0 + 6 ____	Add: ____ = 5 + 3	Add: 8 + 2 ____
G.	**H.**	**I.**
Add: 3 + 3 ____	Add: ____ = 9 + 1	Add: 6 + 7 ____

Level 2b Practice Adding Multi-Digit Numbers
(Adding single digits within 10, horizontally and vertically)

Name: ______________________ Date: ______________________

A. Add: 2 + 1 = ____	**B.** Add: 5 + 4 = ____	**C.** Add: 2 + 6 ____
D. Add: 3 + 4 ____	**E.** Add: ____ = 7 + 0	**F.** Add: 1 + 8 ____
G. Add: 4 + 4 ____	**H.** Add: ____ = 8 + 2	**I.** Add: 7 + 5 ____

Level 1 & 2 Mixed Review Adding Multi-Digit Numbers (Counting on, adding single digits, horizontally and vertically)

Name: ______________________ Date: ______________________

A. Count forward: 21, ___, ___, ___, 25	**B.** Add: 5 + 2 = ___	**C.** Add: $\begin{array}{r} 5 \\ +\,3 \\ \hline \end{array}$
D. Count forward: 45, ___, ___, ___, 49	**E.** Add: ___ = 4 + 4	**F.** Add: $\begin{array}{r} 4 \\ +\,0 \\ \hline \end{array}$
G. Count forward: 78, ___, ___, ___, 82	**H.** Add: ___ = 8 + 2	**I.** Add: $\begin{array}{r} 6 \\ +\,8 \\ \hline \end{array}$

Level 3a Practice Adding Multi-Digit Numbers
(Two-digit + one-digit, no regrouping)

Name: ______________________________ Date: ______________________

A.	**B.**	**C.**
Add: $\begin{array}{r} 12 \\ +\ 3 \\ \hline \end{array}$	Add: $\begin{array}{r} 11 \\ +\ 5 \\ \hline \end{array}$	Add: $\begin{array}{r} 22 \\ +\ 4 \\ \hline \end{array}$
D.	**E.**	**F.**
Add: $\begin{array}{r} 34 \\ +\ 1 \\ \hline \end{array}$	Add: $\begin{array}{r} 27 \\ +\ 2 \\ \hline \end{array}$	Add: $\begin{array}{r} 46 \\ +\ 0 \\ \hline \end{array}$
G.	**H.**	**I.**
Add: $\begin{array}{r} 63 \\ +\ 3 \\ \hline \end{array}$	Add: $\begin{array}{r} 76 \\ +\ 2 \\ \hline \end{array}$	Add: $\begin{array}{r} 96 \\ +\ 3 \\ \hline \end{array}$

Level 3b Practice Adding Multi-Digit Numbers (Two-digit + one-digit, no regrouping)

Name: ______________________________ Date: ______________________

A.	B.	C.
Add: $\begin{array}{r} 11 \\ +4 \\ \hline \end{array}$	Add: $\begin{array}{r} 12 \\ +5 \\ \hline \end{array}$	Add: $\begin{array}{r} 22 \\ +3 \\ \hline \end{array}$
D.	**E.**	**F.**
Add: $\begin{array}{r} 32 \\ +1 \\ \hline \end{array}$	Add: $\begin{array}{r} 26 \\ +2 \\ \hline \end{array}$	Add: $\begin{array}{r} 44 \\ +3 \\ \hline \end{array}$
G.	**H.**	**I.**
Add: $\begin{array}{r} 64 \\ +0 \\ \hline \end{array}$	Add: $\begin{array}{r} 75 \\ +3 \\ \hline \end{array}$	Add: $\begin{array}{r} 97 \\ +2 \\ \hline \end{array}$

Level 4a Practice Adding Multi-Digit Numbers (Two-digit + two-digit, no regrouping)

Name: ______________________________ Date: ______________________

A.	B.	C.
Add: 23 + 15	Add: 11 + 43	Add: 42 + 26
D.	**E.**	**F.**
Add: 32 + 54	Add: 25 + 33	Add: 40 + 23
G.	**H.**	**I.**
Add: 54 + 31	Add: 73 + 14	Add: 64 + 13

Level 4b Practice Adding Multi-Digit Numbers
(Two-digit + two-digit, no regrouping)

Name: ______________________ Date: ______________________

A.	B.	C.
Add: 21 + 14	Add: 13 + 25	Add: 43 + 23
D. Add: 37 + 52	**E.** Add: 35 + 20	**F.** Add: 56 + 12
G. Add: 51 + 35	**H.** Add: 74 + 24	**I.** Add: 84 + 13

Level 3 & 4 Mixed Practice Adding Multi-Digit Numbers (Two-digit + one- or two-digit, no regrouping)

Name: ______________________________ Date: ____________________

A. Add: 32 + 4	**B.** Add: 45 + 1	**C.** Add: 63 + 16
D. Add: 36 + 2	**E.** Add: 15 + 0	**F.** Add: 25 + 52
G. Add: 61 + 33	**H.** Add: 64 + 4	**I.** Add: 21 + 13

Level 5a Practice Adding Multi-Digit Numbers (Two-digit + one-digit, regrouping one's place)

Name: ______________________ Date: ______________

A.	B.	C.
Add: 27 + 5	Add: 11 + 9	Add: 22 + 9
D.	**E.**	**F.**
Add: 36 + 5	Add: 45 + 6	Add: 47 + 6
G.	**H.**	**I.**
Add: 64 + 9	Add: 76 + 8	Add: 67 + 4

Level 5b Practice Adding Multi-Digit Numbers (Two-digit + one-digit, regrouping one's place)

Name: ____________________ Date: ____________________

A. Add: $\begin{array}{r} 27 \\ +4 \\ \hline \end{array}$	**B.** Add: $\begin{array}{r} 12 \\ +8 \\ \hline \end{array}$	**C.** Add: $\begin{array}{r} 23 \\ +9 \\ \hline \end{array}$
D. Add: $\begin{array}{r} 36 \\ +6 \\ \hline \end{array}$	**E.** Add: $\begin{array}{r} 48 \\ +3 \\ \hline \end{array}$	**F.** Add: $\begin{array}{r} 58 \\ +5 \\ \hline \end{array}$
G. Add: $\begin{array}{r} 65 \\ +9 \\ \hline \end{array}$	**H.** Add: $\begin{array}{r} 77 \\ +8 \\ \hline \end{array}$	**I.** Add: $\begin{array}{r} 69 \\ +4 \\ \hline \end{array}$

Level 6a Practice Adding Multi-Digit Numbers (Two-digit + two-digit, regrouping one's place)

Name: ______________________________ Date: ______________________

A. Add: $\begin{array}{r} 27 \\ +24 \\ \hline \end{array}$	**B.** Add: $\begin{array}{r} 11 \\ +39 \\ \hline \end{array}$	**C.** Add: $\begin{array}{r} 22 \\ +49 \\ \hline \end{array}$
D. Add: $\begin{array}{r} 36 \\ +55 \\ \hline \end{array}$	**E.** Add: $\begin{array}{r} 45 \\ +35 \\ \hline \end{array}$	**F.** Add: $\begin{array}{r} 47 \\ +23 \\ \hline \end{array}$
G. Add: $\begin{array}{r} 64 \\ +19 \\ \hline \end{array}$	**H.** Add: $\begin{array}{r} 76 \\ +18 \\ \hline \end{array}$	**I.** Add: $\begin{array}{r} 67 \\ +18 \\ \hline \end{array}$

Level 6b Practice Adding Multi-Digit Numbers (Two-digit + two-digit, regrouping one's place)

Name: ______________________ Date: ______________

A. Add: 26 +25	**B.** Add: 13 +39	**C.** Add: 22 +48
D. Add: 17 +45	**E.** Add: 25 +55	**F.** Add: 46 +24
G. Add: 66 +29	**H.** Add: 54 +18	**I.** Add: 78 +18

Level 5 & 6 Mixed Practice Adding Multi-Digit Numbers (Two-digit + one- or two-digit, regrouping one's place)

Name: ______________________ Date: ______________

A.	B.	C.
Add: $\begin{array}{r} 38 \\ +3 \\ \hline \end{array}$	Add: $\begin{array}{r} 13 \\ +9 \\ \hline \end{array}$	Add: $\begin{array}{r} 19 \\ +22 \\ \hline \end{array}$
D.	**E.**	**F.**
Add: $\begin{array}{r} 17 \\ +43 \\ \hline \end{array}$	Add: $\begin{array}{r} 25 \\ +5 \\ \hline \end{array}$	Add: $\begin{array}{r} 37 \\ +14 \\ \hline \end{array}$
G.	**H.**	**I.**
Add: $\begin{array}{r} 46 \\ +9 \\ \hline \end{array}$	Add: $\begin{array}{r} 66 \\ +26 \\ \hline \end{array}$	Add: $\begin{array}{r} 78 \\ +18 \\ \hline \end{array}$

Level 7a Practice Adding Multi-Digit Numbers
(Two-digit + two-digit, regrouping ten's place, only)

Name: ______________________ Date: ______________

A. Add: 52 +61	**B.** Add: 91 +34	**C.** Add: 81 +41
D. Add: 63 +52	**E.** Add: 64 +71	**F.** Add: 74 +33
G. Add: 64 +90	**H.** Add: 96 +22	**I.** Add: 61 +78

Level 7b Practice Adding Multi-Digit Numbers
(Two-digit + two-digit, regrouping ten's place, only)

Name: ______________________________ Date: ______________________

A.	B.	C.
Add: 42 + 82	Add: 73 + 50	Add: 91 + 33
D.	**E.**	**F.**
Add: 66 + 81	Add: 55 + 51	Add: 64 + 83
G.	**H.**	**I.**
Add: 54 + 92	Add: 94 + 33	Add: 71 + 78

Level 8a Practice Adding Multi-Digit Numbers (Two-digit + two-digit, regrouping BOTH one's and ten's place)

Name: ______________________________ Date: ____________________

A. Add: 83 +28 ___	**B.** Add: 99 +34 ___	**C.** Add: 85 +46 ___
D. Add: 69 +52 ___	**E.** Add: 64 +76 ___	**F.** Add: 74 +37 ___
G. Add: 65 +95 ___	**H.** Add: 96 +25 ___	**I.** Add: 53 +78 ___

Level 8b Practice Adding Multi-Digit Numbers (Two-digit + two-digit, regrouping BOTH one's and ten's place)

Name: ______________________________ Date: ______________________

A.	B.	C.
Add: $\begin{array}{r} 44 \\ +87 \\ \hline \end{array}$	Add: $\begin{array}{r} 99 \\ +21 \\ \hline \end{array}$	Add: $\begin{array}{r} 76 \\ +45 \\ \hline \end{array}$
D.	**E.**	**F.**
Add: $\begin{array}{r} 69 \\ +84 \\ \hline \end{array}$	Add: $\begin{array}{r} 64 \\ +46 \\ \hline \end{array}$	Add: $\begin{array}{r} 56 \\ +47 \\ \hline \end{array}$
G.	**H.**	**I.**
Add: $\begin{array}{r} 15 \\ +95 \\ \hline \end{array}$	Add: $\begin{array}{r} 96 \\ +26 \\ \hline \end{array}$	Add: $\begin{array}{r} 84 \\ +78 \\ \hline \end{array}$

Level 7 & 8 Mixed Practice Adding Multi-Digit Numbers (Two-digit + two-digit, regrouping one's or regrouping BOTH one's and ten's place)

Name: ______________________ Date: ______________________

A.

Add:

$$\begin{array}{r} 27 \\ +14 \\ \hline \end{array}$$

B.

Add:

$$\begin{array}{r} 69 \\ +22 \\ \hline \end{array}$$

C.

Add:

$$\begin{array}{r} 66 \\ +56 \\ \hline \end{array}$$

D.

Add:

$$\begin{array}{r} 58 \\ +23 \\ \hline \end{array}$$

E.

Add:

$$\begin{array}{r} 68 \\ +32 \\ \hline \end{array}$$

F.

Add:

$$\begin{array}{r} 63 \\ +59 \\ \hline \end{array}$$

G.

Add:

$$\begin{array}{r} 65 \\ +47 \\ \hline \end{array}$$

H.

Add:

$$\begin{array}{r} 87 \\ +46 \\ \hline \end{array}$$

I.

Add:

$$\begin{array}{r} 74 \\ +77 \\ \hline \end{array}$$

Level 9a Practice Adding Multi-Digit Numbers (Three-digit + one-digit, with and without regrouping)

Name: ______________________ Date: ______________________

A.	B.	C.
Add: $\begin{array}{r} 253 \\ +\quad 2 \\ \hline \end{array}$	Add: $\begin{array}{r} 341 \\ +\quad 4 \\ \hline \end{array}$	Add: $\begin{array}{r} 585 \\ +\quad 3 \\ \hline \end{array}$
D.	**E.**	**F.**
Add: $\begin{array}{r} 460 \\ +\quad 2 \\ \hline \end{array}$	Add: $\begin{array}{r} 664 \\ +\quad 6 \\ \hline \end{array}$	Add: $\begin{array}{r} 474 \\ +\quad 7 \\ \hline \end{array}$
G.	**H.**	**I.**
Add: $\begin{array}{r} 365 \\ +\quad 6 \\ \hline \end{array}$	Add: $\begin{array}{r} 459 \\ +\quad 1 \\ \hline \end{array}$	Add: $\begin{array}{r} 823 \\ +\quad 8 \\ \hline \end{array}$

Level 9b Practice Adding Multi-Digit Numbers (Three-digit + one-digit, with and without regrouping)

Name: ______________________________ Date: ____________________

A.	B.	C.
Add: 261 + 4	Add: 150 + 6	Add: 482 + 2
D.	**E.**	**F.**
Add: 576 + 2	Add: 617 + 9	Add: 174 + 8
G.	**H.**	**I.**
Add: 365 + 2	Add: 449 + 2	Add: 935 + 8

Level 10a Practice Adding Multi-Digit Numbers (Three-digit + two-digit, with and without regrouping one place value)

Name: ______________________________ Date: ______________________

A.	B.	C.
Add: 453 + 22	Add: 201 + 34	Add: 185 + 10

D.	E.	F.
Add: 566 + 15	Add: 264 + 27	Add: 364 + 43

G.	H.	I.
Add: 265 + 91	Add: 346 + 15	Add: 413 + 78

Level 10b Practice Adding Multi-Digit Numbers (Three-digit + two-digit, with and without regrouping one place value)

Name: ______________________ Date: ______________________

A. Add: 211 + 32	**B.** Add: 132 + 24	**C.** Add: 275 + 24
D. Add: 306 + 28	**E.** Add: 554 + 18	**F.** Add: 461 + 60
G. Add: 265 + 83	**H.** Add: 790 + 47	**I.** Add: 204 + 58

Level 9 & 10 Practice Adding Multi-Digit Numbers (Three-digit + one- or two-digit, with and without regrouping one place value)

Name: ______________________ Date: ______________

A.	B.	C.
Add: 232 + 2	Add: 351 + 14	Add: 225 + 16

D.	E.	F.
Add: 346 + 28	Add: 274 + 8	Add: 461 + 63

G.	H.	I.
Add: 165 + 91	Add: 435 + 7	Add: 374 + 52

Level 11a Practice Adding Multi-Digit Numbers (Three-digit + three-digit, no regrouping)

Name: ____________________ Date: ____________________

A. Add: 243 + 123	**B.** Add: 352 + 423	**C.** Add: 743 + 222
D. Add: 123 + 321	**E.** Add: 468 + 311	**F.** Add: 544 + 124
G. Add: 643 + 126	**H.** Add: 213 + 120	**I.** Add: 2,243 + 4,123

Level 11b Practice Adding Multi-Digit Numbers (Three-digit + three-digit, no regrouping)

Name: ______________________ Date: ______________

A.	B.	C.
Add: $\begin{array}{r} 123 \\ +\,343 \\ \hline \end{array}$	Add: $\begin{array}{r} 334 \\ +\,523 \\ \hline \end{array}$	Add: $\begin{array}{r} 653 \\ +\,112 \\ \hline \end{array}$
D.	**E.**	**F.**
Add: $\begin{array}{r} 412 \\ +\,475 \\ \hline \end{array}$	Add: $\begin{array}{r} 466 \\ +\,412 \\ \hline \end{array}$	Add: $\begin{array}{r} 521 \\ +\,326 \\ \hline \end{array}$
G.	**H.**	**I.**
Add: $\begin{array}{r} 763 \\ +\,125 \\ \hline \end{array}$	Add: $\begin{array}{r} 243 \\ +\,320 \\ \hline \end{array}$	Add: $\begin{array}{r} 1{,}147 \\ +\,5{,}632 \\ \hline \end{array}$

Level 12a Practice Adding Multi-Digit Numbers
(Three-digit + three-digit, regrouping one's or ten's place)

Name: ______________________________ Date: ______________________

A.	B.	C.
Add: 243 +128	Add: 352 +429	Add: 745 +125
D. Add: 127 +324	**E.** Add: 468 +313	**F.** Add: 544 +194
G. Add: 643 +186	**H.** Add: 263 +160	**I.** Add: 2,243 +4,128

Level 12b Practice Adding Multi-Digit Numbers
(Three-digit + three-digit, regrouping one's or ten's place)

Name: ______________________ Date: ______________________

A.	B.	C.
Add: 314 + 128	Add: 335 + 126	Add: 646 + 214
D.	**E.**	**F.**
Add: 227 + 455	Add: 765 + 171	Add: 544 + 182
G.	**H.**	**I.**
Add: 442 + 196	Add: 264 + 260	Add: 2,343 + 6,219

Level 11 & 12 Mixed Practice Adding Multi-Digit Numbers (Three-digit + three-digit, no regrouping or regrouping in one place value)

Name: ______________________ Date: ______________________

A. Add: 221 + 135	**B.** Add: 394 + 523	**C.** Add: 423 + 217
D. Add: 233 + 593	**E.** Add: 162 + 315	**F.** Add: 128 + 336
G. Add: 763 + 154	**H.** Add: 327 + 430	**I.** Add: 1,147 + 5,636

Level 13a Practice Adding Multi-Digit Numbers (Three-digit + three-digit, regrouping two place values)

Name: ______________________ Date: ______________________

A.	B.	C.
Add: $\begin{array}{r} 243 \\ +178 \\ \hline \end{array}$	Add: $\begin{array}{r} 352 \\ +459 \\ \hline \end{array}$	Add: $\begin{array}{r} 745 \\ +185 \\ \hline \end{array}$
D.	**E.**	**F.**
Add: $\begin{array}{r} 187 \\ +374 \\ \hline \end{array}$	Add: $\begin{array}{r} 468 \\ +353 \\ \hline \end{array}$	Add: $\begin{array}{r} 544 \\ +199 \\ \hline \end{array}$
G.	**H.**	**I.**
Add: $\begin{array}{r} 648 \\ +186 \\ \hline \end{array}$	Add: $\begin{array}{r} 264 \\ +168 \\ \hline \end{array}$	Add: $\begin{array}{r} 2,243 \\ +4,188 \\ \hline \end{array}$

Level 13b Practice Adding Multi-Digit Numbers
(Three-digit + three-digit, regrouping two place values)

Name: ______________________ Date: ______________________

A.	B.	C.
Add: $\begin{array}{r} 154 \\ +178 \\ \hline \end{array}$	Add: $\begin{array}{r} 263 \\ +447 \\ \hline \end{array}$	Add: $\begin{array}{r} 555 \\ +286 \\ \hline \end{array}$
D.	**E.**	**F.**
Add: $\begin{array}{r} 187 \\ +394 \\ \hline \end{array}$	Add: $\begin{array}{r} 169 \\ +263 \\ \hline \end{array}$	Add: $\begin{array}{r} 523 \\ +299 \\ \hline \end{array}$
G.	**H.**	**I.**
Add: $\begin{array}{r} 348 \\ +168 \\ \hline \end{array}$	Add: $\begin{array}{r} 267 \\ +457 \\ \hline \end{array}$	Add: $\begin{array}{r} 3,155 \\ +3,188 \\ \hline \end{array}$

Level 14a Practice Adding Multi-Digit Numbers
(Three- or four-digit + three-digit, regrouping into the thousand's place)

Name: ______________________________ Date: ______________________

A. Add: 243 + 935	**B.** Add: 812 + 451	**C.** Add: 745 + 514
D. Add: 3,983 + 314	**E.** Add: 1,462 + 623	**F.** Add: 5,544 + 622
G. Add: 2,623 + 464	**H.** Add: 252 + 802	**I.** Add: 2,203 + 981

Level 14b Practice Adding Multi-Digit Numbers
(Three- or four-digit + three-digit, regrouping into the thousand's place)

Name: ____________________________ Date: ____________________

A. Add: 312 + 814	**B.** Add: 712 + 541	**C.** Add: 942 + 424
D. Add: 3,663 + 511	**E.** Add: 4,562 + 511	**F.** Add: 5,621 + 630
G. Add: 2,703 + 555	**H.** Add: 552 + 821	**I.** Add: 3,732 + 715

Level 13 & 14 Mixed Practice Adding Multi-Digit Numbers (Three-digit + three-digit, regrouping two place values or regrouping into the thousand's place)

Name: ______________________________ Date: ______________________

A. Add: 254 +357	**B.** Add: 634 +412	**C.** Add: 436 +298
D. Add: 910 +524	**E.** Add: 3,856 + 711	**F.** Add: 1,535 + 187
G. Add: 3,424 + 921	**H.** Add: 6,656 + 463	**I.** Add: 2,243 +4,188

Addition Trajectory Answer Keys

Level 1a Practice Adding Multi-Digit Numbers
(Counting on within 100)
ANSWER KEY

Name: ______________________ Date: ______________

A.	B.	C.
Count forward: 4, 5, 6, 7, 8	Count forward: 12, 13, 14, 15, 16	Count forward: 21, 22, 23, 24, 25
D. Count forward: 45, 46, 47, 48, 49	**E.** Count forward: 68, 69, 70, 71, 72	**F.** Count forward: 99, 100, 101, 102, 103

Level 1b Practice Adding Multi-Digit Numbers
(Counting on within 100)
ANSWER KEY

Name: ______________________ Date: ______________________

A. Count forward: 2, 3, 4, 5, 6	**B.** Count forward: 10, 11, 12, 13, 14	**C.** Count forward: 23, 24, 25, 26, 27
D. Count forward: 41, 42, 43, 44, 45	**E.** Count forward: 67, 68, 69, 70, 71	**F.** Count forward: 98, 99, 100, 101, 102

Level 2a Practice Adding Multi-Digit Numbers
(Adding single digits within 10, horizontally and vertically)
ANSWER KEY

Name: ______________________________ Date: ______________________

A.	B.	C.
Add: 3 + 1 = 4	Add: 4 + 3 = 7	Add: 2 + 7 ――― 9
D.	**E.**	**F.**
Add: 0 + 6 ――― 6	Add: 8 = 5 + 3	Add: 8 + 2 ――― 10
G.	**H.**	**I.**
Add: 3 + 3 ――― 6	Add: 10 = 9 + 1	Add: 6 + 7 ――― 13

Level 2b Practice Adding Multi-Digit Numbers
(Adding single digits within 10, horizontally and vertically)
ANSWER KEY

Name: ______________________________ Date: ____________________

A.	B.	C.
Add: $2 + 1 = \underline{3}$	Add: $5 + 4 = \underline{9}$	Add: $\begin{array}{r} 2 \\ +\,6 \\ \hline 8 \end{array}$
D.	**E.**	**F.**
Add: $\begin{array}{r} 3 \\ +\,4 \\ \hline 7 \end{array}$	Add: $\underline{7} = 7 + 0$	Add: $\begin{array}{r} 1 \\ +\,8 \\ \hline 9 \end{array}$
G.	**H.**	**I.**
Add: $\begin{array}{r} 4 \\ +\,4 \\ \hline 8 \end{array}$	Add: $\underline{10} = 8 + 2$	Add: $\begin{array}{r} 7 \\ +\,5 \\ \hline 12 \end{array}$

Level 1 & 2 Mixed Review Adding Multi-Digit Numbers (Counting on, Adding single digits, horizontally and vertically) ANSWER KEY

Name: ______________________ Date: ______________________

A. Count forward: 21, 22, 23, 24, 25	**B.** Add: 5 + 2 = 7	**C.** Add: $\begin{array}{r} 5 \\ +3 \\ \hline 8 \end{array}$
D. Count forward: 45, 46, 47, 48, 49	**E.** Add: 8 = 4 + 4	**F.** Add: $\begin{array}{r} 4 \\ +0 \\ \hline 4 \end{array}$
G. Count forward: 78, 79, 80, 81, 82	**H.** Add: 10 = 8 + 2	**I.** Add: $\begin{array}{r} 6 \\ +8 \\ \hline 14 \end{array}$

Level 3a Practice Adding Multi-Digit Numbers
(Two-digit + one-digit, no regrouping)
ANSWER KEY

Name: ______________________ Date: ______________________

A.	B.	C.
Add: 12 + 3 = 15	Add: 11 + 5 = 16	Add: 22 + 4 = 26

D.	E.	F.
Add: 34 + 1 = 35	Add: 27 + 2 = 29	Add: 46 + 0 = 46

G.	H.	I.
Add: 63 + 3 = 66	Add: 76 + 2 = 78	Add: 96 + 3 = 99

Level 3b Practice Adding Multi-Digit Numbers
(Two-digit + one-digit, no regrouping)
ANSWER KEY

Name: ______________________ Date: ______________________

A.	B.	C.
Add: $\begin{array}{r} 11 \\ +4 \\ \hline 15 \end{array}$	Add: $\begin{array}{r} 12 \\ +5 \\ \hline 17 \end{array}$	Add: $\begin{array}{r} 22 \\ +3 \\ \hline 25 \end{array}$
D. Add: $\begin{array}{r} 32 \\ +1 \\ \hline 33 \end{array}$	**E.** Add: $\begin{array}{r} 26 \\ +2 \\ \hline 28 \end{array}$	**F.** Add: $\begin{array}{r} 44 \\ +3 \\ \hline 47 \end{array}$
G. Add: $\begin{array}{r} 64 \\ +0 \\ \hline 64 \end{array}$	**H.** Add: $\begin{array}{r} 75 \\ +3 \\ \hline 78 \end{array}$	**I.** Add: $\begin{array}{r} 97 \\ +2 \\ \hline 99 \end{array}$

Level 4a Practice Adding Multi-Digit Numbers (Two-digit + two-digit, no regrouping) ANSWER KEY

Name: ______________________ Date: ______________________

A.	B.	C.
Add: $\begin{array}{r} 23 \\ +15 \\ \hline 38 \end{array}$	Add: $\begin{array}{r} 11 \\ +43 \\ \hline 54 \end{array}$	Add: $\begin{array}{r} 42 \\ +26 \\ \hline 68 \end{array}$
D.	**E.**	**F.**
Add: $\begin{array}{r} 32 \\ +54 \\ \hline 86 \end{array}$	Add: $\begin{array}{r} 25 \\ +33 \\ \hline 58 \end{array}$	Add: $\begin{array}{r} 40 \\ +23 \\ \hline 63 \end{array}$
G.	**H.**	**I.**
Add: $\begin{array}{r} 54 \\ +31 \\ \hline 85 \end{array}$	Add: $\begin{array}{r} 73 \\ +14 \\ \hline 87 \end{array}$	Add: $\begin{array}{r} 64 \\ +13 \\ \hline 77 \end{array}$

Level 4b Practice Adding Multi-Digit Numbers
(Two-digit + two-digit, no regrouping)
ANSWER KEY

Name: ______________________ Date: ______________

A. Add: $\begin{array}{r} 21 \\ +14 \\ \hline 35 \end{array}$	**B.** Add: $\begin{array}{r} 13 \\ +25 \\ \hline 38 \end{array}$	**C.** Add: $\begin{array}{r} 43 \\ +23 \\ \hline 66 \end{array}$
D. Add: $\begin{array}{r} 37 \\ +52 \\ \hline 89 \end{array}$	**E.** Add: $\begin{array}{r} 35 \\ +20 \\ \hline 55 \end{array}$	**F.** Add: $\begin{array}{r} 56 \\ +12 \\ \hline 68 \end{array}$
G. Add: $\begin{array}{r} 51 \\ +35 \\ \hline 86 \end{array}$	**H.** Add: $\begin{array}{r} 74 \\ +24 \\ \hline 98 \end{array}$	**I.** Add: $\begin{array}{r} 84 \\ +13 \\ \hline 97 \end{array}$

Level 3 & 4 Mixed Practice Adding Multi-Digit Numbers
(Two-digit + one- or two-digit, no regrouping)
ANSWER KEY

Name: ____________________ Date: ____________________

A.	B.	C.
Add: 32 + 4 = 36	Add: 45 + 1 = 46	Add: 63 + 16 = 79

D.	E.	F.
Add: 36 + 2 = 38	Add: 15 + 0 = 15	Add: 25 + 52 = 77

G.	H.	I.
Add: 61 + 33 = 94	Add: 64 + 4 = 68	Add: 21 + 13 = 34

Level 5a Practice Adding Multi-Digit Numbers (Two-digit + one-digit, regrouping one's place) ANSWER KEY

Name: ______________________ Date: ______________________

A.	B.	C.
Add: 27 + 5 = 32	Add: 11 + 9 = 20	Add: 22 + 9 = 31
D.	**E.**	**F.**
Add: 36 + 5 = 41	Add: 45 + 6 = 51	Add: 47 + 6 = 53
G.	**H.**	**I.**
Add: 64 + 9 = 73	Add: 76 + 8 = 84	Add: 67 + 4 = 71

Level 5b Practice Adding Multi-Digit Numbers (Two-digit + one-digit, regrouping one's place)
ANSWER KEY

Name: ______________________ Date: ______________________

A. Add: 27 + 4 31	**B.** Add: 12 + 8 20	**C.** Add: 23 + 9 32
D. Add: 36 + 6 42	**E.** Add: 48 + 3 51	**F.** Add: 58 + 5 63
G. Add: 65 + 9 74	**H.** Add: 77 + 8 85	**I.** Add: 69 + 4 73

Level 6a Practice Adding Multi-Digit Numbers
(Two-digit + two-digit, regrouping one's place)
ANSWER KEY

Name: ______________________________ Date: ______________________

A.	B.	C.
Add: 27 + 24 = 51	Add: 11 + 39 = 50	Add: 22 + 49 = 71
D.	**E.**	**F.**
Add: 36 + 55 = 91	Add: 45 + 35 = 80	Add: 47 + 23 = 70
G.	**H.**	**I.**
Add: 64 + 19 = 83	Add: 76 + 18 = 94	Add: 67 + 18 = 85

Level 6b Practice Adding Multi-Digit Numbers
(Two-digit + two-digit, regrouping one's place)
ANSWER KEY

Name: ______________________ Date: ______________

A.
Add:
$$\begin{array}{r} 26 \\ +25 \\ \hline 51 \end{array}$$

B.
Add:
$$\begin{array}{r} 13 \\ +39 \\ \hline 52 \end{array}$$

C.
Add:
$$\begin{array}{r} 22 \\ +48 \\ \hline 70 \end{array}$$

D.
Add:
$$\begin{array}{r} 17 \\ +45 \\ \hline 62 \end{array}$$

E.
Add:
$$\begin{array}{r} 25 \\ +55 \\ \hline 80 \end{array}$$

F.
Add:
$$\begin{array}{r} 46 \\ +24 \\ \hline 70 \end{array}$$

G.
Add:
$$\begin{array}{r} 66 \\ +29 \\ \hline 95 \end{array}$$

H.
Add:
$$\begin{array}{r} 54 \\ +18 \\ \hline 72 \end{array}$$

I.
Add:
$$\begin{array}{r} 78 \\ +18 \\ \hline 96 \end{array}$$

Level 5 & 6 Mixed Practice Adding Multi-Digit Numbers
(Two-digit + one- or two-digit, regrouping one's place)
ANSWER KEY

Name: ______________________ Date: ______________________

A.	B.	C.
Add: $\begin{array}{r} 38 \\ +3 \\ \hline 41 \end{array}$	Add: $\begin{array}{r} 13 \\ +9 \\ \hline 22 \end{array}$	Add: $\begin{array}{r} 19 \\ +22 \\ \hline 41 \end{array}$
D.	**E.**	**F.**
Add: $\begin{array}{r} 17 \\ +43 \\ \hline 60 \end{array}$	Add: $\begin{array}{r} 25 \\ +5 \\ \hline 30 \end{array}$	Add: $\begin{array}{r} 37 \\ +14 \\ \hline 51 \end{array}$
G.	**H.**	**I.**
Add: $\begin{array}{r} 46 \\ +9 \\ \hline 55 \end{array}$	Add: $\begin{array}{r} 66 \\ +26 \\ \hline 92 \end{array}$	Add: $\begin{array}{r} 78 \\ +18 \\ \hline 96 \end{array}$

Level 7a Practice Adding Multi-Digit Numbers
(Two-digit + two-digit, regrouping ten's place, only)
ANSWER KEY

Name: ______________________ Date: ______________

A.	B.	C.
Add: 52 + 61 = 113	Add: 91 + 34 = 125	Add: 81 + 41 = 122

D.	E.	F.
Add: 63 + 52 = 115	Add: 64 + 71 = 135	Add: 74 + 33 = 107

G.	H.	I.
Add: 64 + 90 = 154	Add: 96 + 22 = 118	Add: 61 + 78 = 139

Level 7b Practice Adding Multi-Digit Numbers (Two-digit + two-digit, regrouping ten's place, only) ANSWER KEY

Name: ______________________ Date: ______________________

A.	B.	C.
Add: $\begin{array}{r} 42 \\ +82 \\ \hline 124 \end{array}$	Add: $\begin{array}{r} 73 \\ +50 \\ \hline 123 \end{array}$	Add: $\begin{array}{r} 91 \\ +33 \\ \hline 124 \end{array}$
D.	**E.**	**F.**
Add: $\begin{array}{r} 66 \\ +81 \\ \hline 147 \end{array}$	Add: $\begin{array}{r} 55 \\ +51 \\ \hline 106 \end{array}$	Add: $\begin{array}{r} 64 \\ +83 \\ \hline 147 \end{array}$
G.	**H.**	**I.**
Add: $\begin{array}{r} 54 \\ +92 \\ \hline 146 \end{array}$	Add: $\begin{array}{r} 94 \\ +33 \\ \hline 127 \end{array}$	Add: $\begin{array}{r} 71 \\ +78 \\ \hline 149 \end{array}$

Level 8a Practice Adding Multi-Digit Numbers
(Two-digit + two-digit, regrouping BOTH one's and ten's place)
ANSWER KEY

Name: ______________________ Date: ______________________

A.	B.	C.
Add: 83 + 28 = 111	Add: 99 + 34 = 133	Add: 85 + 46 = 131

D.	E.	F.
Add: 69 + 52 = 121	Add: 64 + 76 = 140	Add: 74 + 37 = 111

G.	H.	I.
Add: 65 + 95 = 160	Add: 96 + 25 = 121	Add: 53 + 78 = 131

Level 8b Practice Adding Multi-Digit Numbers (Two-digit + two-digit, regrouping BOTH one's and ten's place) ANSWER KEY

Name: ______________________________ Date: ______________________

A.	B.	C.
Add: 44 + 87 = 131	Add: 99 + 21 = 120	Add: 76 + 45 = 121
D.	**E.**	**F.**
Add: 69 + 84 = 153	Add: 64 + 46 = 110	Add: 56 + 47 = 103
G.	**H.**	**I.**
Add: 15 + 95 = 110	Add: 96 + 26 = 122	Add: 84 + 78 = 162

Level 7 & 8 Mixed Practice Adding Multi-Digit Numbers (Two-digit + two-digit, regrouping one's or regrouping BOTH one's and ten's place)
ANSWER KEY

Name: ______________________ Date: ______________________

A.	B.	C.
Add: 27 + 14 = 41	Add: 69 + 22 = 91	Add: 66 + 56 = 122

D.	E.	F.
Add: 58 + 23 = 81	Add: 68 + 32 = 100	Add: 63 + 59 = 122

G.	H.	I.
Add: 65 + 47 = 112	Add: 87 + 46 = 133	Add: 74 + 77 = 151

Level 9a Practice Adding Multi-Digit Numbers
(Three-digit + one-digit, with and without regrouping)
ANSWER KEY

Name: ______________________ Date: ______________

A.	**B.**	**C.**
Add: 253 + 2 = 255	Add: 341 + 4 = 345	Add: 585 + 3 = 588
D.	**E.**	**F.**
Add: 460 + 2 = 462	Add: 664 + 6 = 670	Add: 474 + 7 = 481
G.	**H.**	**I.**
Add: 365 + 6 = 371	Add: 459 + 1 = 460	Add: 823 + 8 = 831

Level 9b Practice Adding Multi-Digit Numbers
(Three-digit + one-digit, with and without regrouping)
ANSWER KEY

Name: ______________________ Date: ______________________

A.	B.	C.
Add: 261 + 4 = 265	Add: 150 + 6 = 156	Add: 482 + 2 = 484

D.	E.	F.
Add: 576 + 2 = 578	Add: 617 + 9 = 626	Add: 174 + 8 = 182

G.	H.	I.
Add: 365 + 2 = 367	Add: 449 + 2 = 451	Add: 935 + 8 = 943

Level 10a Practice Adding Multi-Digit Numbers
(Three-digit + two-digit, with and without regrouping one place value)
ANSWER KEY

Name: ______________________ Date: ______________________

A.	B.	C.
Add: 453 + 22 = 475	Add: 201 + 34 = 235	Add: 185 + 10 = 195

D.	E.	F.
Add: 566 + 15 = 581	Add: 264 + 27 = 291	Add: 364 + 43 = 407

G.	H.	I.
Add: 265 + 91 = 356	Add: 346 + 15 = 361	Add: 413 + 78 = 491

Level 10b Practice Adding Multi-Digit Numbers
(Three-digit + two-digit, with and without regrouping one place value)
ANSWER KEY

Name: ______________________ Date: ______________

A.	**B.**	**C.**
Add: 211 + 32 = 243	Add: 132 + 24 = 156	Add: 275 + 24 = 299
D.	**E.**	**F.**
Add: 306 + 28 = 334	Add: 554 + 18 = 572	Add: 461 + 60 = 521
G.	**H.**	**I.**
Add: 265 + 83 = 348	Add: 790 + 47 = 837	Add: 204 + 58 = 262

Level 9 & 10 Practice Adding Multi-Digit Numbers (Three-digit + one- or two-digit, with and without regrouping one place value)
ANSWER KEY

Name: ______________________ Date: ______________________

A.	B.	C.
Add: 232 + 2 = 234	Add: 351 + 14 = 365	Add: 225 + 16 = 241
D.	**E.**	**F.**
Add: 346 + 28 = 374	Add: 274 + 8 = 282	Add: 461 + 63 = 524
G.	**H.**	**I.**
Add: 165 + 91 = 256	Add: 435 + 7 = 442	Add: 374 + 52 = 426

Level 11a Practice Adding Multi-Digit Numbers (Three-digit + three-digit, no regrouping) ANSWER KEY

Name: ______________________ Date: ______________________

A.	B.	C.
Add: 243 + 123 = 366	Add: 352 + 423 = 775	Add: 743 + 222 = 965
D.	**E.**	**F.**
Add: 123 + 321 = 444	Add: 468 + 311 = 779	Add: 544 + 124 = 668
G.	**H.**	**I.**
Add: 643 + 126 = 769	Add: 213 + 120 = 333	Add: 2,243 + 4,123 = 6,366

Level 11b Practice Adding Multi-Digit Numbers (Three-digit + three-digit, no regrouping)
ANSWER KEY

Name: ______________________________ Date: ______________________

A. Add: 123 +343 466	**B.** Add: 334 +523 857	**C.** Add: 653 +112 765
D. Add: 412 +475 887	**E.** Add: 466 +412 878	**F.** Add: 521 +326 847
G. Add: 763 +125 888	**H.** Add: 243 +320 563	**I.** Add: 1,147 +5,632 6,779

Level 12a Practice Adding Multi-Digit Numbers
(Three-digit + three-digit, regrouping one's or ten's place)
ANSWER KEY

Name: ______________________ Date: ______________________

A.	B.	C.
Add: 243 + 128 = 371	Add: 352 + 429 = 781	Add: 745 + 125 = 870

D.	E.	F.
Add: 127 + 324 = 451	Add: 468 + 313 = 781	Add: 544 + 194 = 738

G.	H.	I.
Add: 643 + 186 = 829	Add: 263 + 160 = 423	Add: 2,243 + 4,128 = 6,371

Level 12b Practice Adding Multi-Digit Numbers
(Three-digit + three-digit, regrouping one's or ten's place)
ANSWER KEY

Name: ______________________________ Date: ______________________

A.	B.	C.
Add: $\begin{array}{r} 314 \\ +128 \\ \hline 442 \end{array}$	Add: $\begin{array}{r} 335 \\ +126 \\ \hline 461 \end{array}$	Add: $\begin{array}{r} 646 \\ +214 \\ \hline 860 \end{array}$
D.	**E.**	**F.**
Add: $\begin{array}{r} 227 \\ +455 \\ \hline 682 \end{array}$	Add: $\begin{array}{r} 765 \\ +171 \\ \hline 936 \end{array}$	Add: $\begin{array}{r} 544 \\ +182 \\ \hline 726 \end{array}$
G.	**H.**	**I.**
Add: $\begin{array}{r} 442 \\ +196 \\ \hline 638 \end{array}$	Add: $\begin{array}{r} 264 \\ +260 \\ \hline 524 \end{array}$	Add: $\begin{array}{r} 2,343 \\ +6,219 \\ \hline 8,562 \end{array}$

Level 11 & 12 Mixed Practice Adding Multi-Digit Numbers (Three-digit + three-digit, no regrouping or regrouping in one's or ten's place) ANSWER KEY

Name: ______________________ Date: ______________

A.	B.	C.
Add: 221 + 135 = 356	Add: 394 + 523 = 917	Add: 423 + 217 = 640

D.	E.	F.
Add: 233 + 593 = 826	Add: 162 + 315 = 477	Add: 128 + 336 = 464

G.	H.	I.
Add: 763 + 154 = 917	Add: 327 + 430 = 757	Add: 1,147 + 5,636 = 6,783

Level 13a Practice Adding Multi-Digit Numbers
(Three-digit + three-digit, regrouping two place values)
ANSWER KEY

Name: ______________________ Date: ______________________

A.	B.	C.
Add: 243 + 178 = 421	Add: 352 + 459 = 811	Add: 745 + 185 = 930

D.	E.	F.
Add: 187 + 374 = 561	Add: 468 + 353 = 821	Add: 544 + 199 = 743

G.	H.	I.
Add: 648 + 186 = 834	Add: 264 + 168 = 432	Add: 2, 243 + 4, 188 = 6,431

Level 13b Practice Adding Multi-Digit Numbers (Three-digit + three-digit, regrouping two place values) ANSWER KEY

Name: ______________________________ Date: ____________________

A.

Add:

$$\begin{array}{r} 154 \\ +\,178 \\ \hline 332 \end{array}$$

B.

Add:

$$\begin{array}{r} 263 \\ +\,447 \\ \hline 710 \end{array}$$

C.

Add:

$$\begin{array}{r} 555 \\ +\,286 \\ \hline 841 \end{array}$$

D.

Add:

$$\begin{array}{r} 187 \\ +\,394 \\ \hline 581 \end{array}$$

E.

Add:

$$\begin{array}{r} 169 \\ +\,263 \\ \hline 432 \end{array}$$

F.

Add:

$$\begin{array}{r} 523 \\ +\,299 \\ \hline 822 \end{array}$$

G.

Add:

$$\begin{array}{r} 348 \\ +\,168 \\ \hline 516 \end{array}$$

H.

Add:

$$\begin{array}{r} 267 \\ +\,457 \\ \hline 724 \end{array}$$

I.

Add:

$$\begin{array}{r} 3{,}155 \\ +\,3{,}188 \\ \hline 6{,}343 \end{array}$$

Level 14a Practice Adding Multi-Digit Numbers (Three- or four-digit + three-digit, regrouping into the thousand's place) ANSWER KEY

Name: ______________________ Date: ______________

A.	B.	C.
Add: 243 + 935 = 1,178	Add: 812 + 451 = 1,263	Add: 745 + 514 = 1,259
D.	**E.**	**F.**
Add: 3,983 + 314 = 4,297	Add: 1,462 + 623 = 2,085	Add: 5,544 + 622 = 6,166
G.	**H.**	**I.**
Add: 2,623 + 464 = 3,087	Add: 252 + 802 = 1,054	Add: 2,203 + 981 = 3,184

Level 14b Practice Adding Multi-Digit Numbers (Three- or four-digit + three-digit, regrouping into the thousand's place) ANSWER KEY

Name: ______________________________ Date: ______________________

A.	B.	C.
Add: 312 +814 1,126	Add: 712 +541 1,253	Add: 942 +424 1,366

D.	E.	F.
Add: 3,663 + 511 4,174	Add: 4,562 + 511 5,073	Add: 5,621 + 630 6,251

G.	H.	I.
Add: 2,703 + 555 3,258	Add: 552 +821 1,373	Add: 3,732 + 715 4,447

Level 13 & 14 Mixed Practice Adding Multi-Digit Numbers (Three-digit + three-digit, regrouping two place values or regrouping into the thousand's place)
ANSWER KEY

Name: ______________________ Date: ______________________

A.	B.	C.
Add: 254 + 357 = 611	Add: 634 + 412 = 1,046	Add: 436 + 298 = 734
D.	**E.**	**F.**
Add: 910 + 524 = 1,434	Add: 3,856 + 711 = 4,567	Add: 1,535 + 187 = 1,722
G.	**H.**	**I.**
Add: 3,424 + 921 = 4,345	Add: 6,656 + 463 = 7,119	Add: 2,243 + 4,188 = 6,431

Subtraction Trajectory

Subtraction Skills Trajectory
Multi-Digit Subtraction for Whole Numbers

(Levels on the trajectory match the numbers on the Multi-Digit Subtraction Pre- and Post-Assessment)

First Grade begins here

Assumptions

- ✔ Understanding of the concept of subtraction (taking away)
- ✔ Being able to show subtraction with manipulatives as taking away

Level 1 – Counting back from any number 0-100.

Level 2 – Subtracting single digits both horizontally and vertically with or without manipulatives (fingers are fine).

Second grade begins here

Level 3 – Two-digit minus one digit, no regrouping
Level 4 – Two-digit minus two-digit, no regrouping
Level 5 – Two-digit minus one-digit, regrouping one's place
Level 6 – Two-digit minus two-digit, regrouping one's place
Level 7 – Three-digit minus one-digit, with and without regrouping
Level 8 – Three-digit minus two-digit, regrouping ten's place
Level 9 – Three-digit minus two-digit, regrouping two place values
Level 10 – Three-digit minus three-digit, no regrouping
Level 11 – Three-digit minus three-digit, regrouping one place value
Level 12 – Three-digit minus three-digit, regrouping two place values
Level 13 – Regrouping across one zero
Level 14 – Regrouping across two zeros

Pre and Post Assessment for Subtracting Multi-Digit Numbers

Name: ______________________________ Date: ________________

1. Count backwards:	**2.** Subtract:	**3.** Subtract:
___, ___, ___, ___, 63	9 – 3 = ___	14 – 3
4. Subtract:	**5.** Subtract:	**6.** Subtract:
54 – 21	25 – 7	44 – 27
7. Subtract:	**8.** Subtract:	**9.** Subtract:
365 – 7	325 – 81	750 – 83

Pre and Post Assessment for Subtracting Multi-Digit Numbers

Name: ______________________________ Date: ____________________

10. Subtract: 456 − 123	**11.** Subtract: 651 − 114	**12.** Subtract: 635 − 387
13. Subtract: 403 − 27	**14.** Subtract: 200 − 68	

Pre and Post Assessment for Subtracting Multi-Digit Numbers
ANSWER KEY

Name: ____________________ Date: ____________

1. Count backwards: 59, 60, 61, 62, 63	**2.** Subtract: $9 - 3 = \underline{6}$	**3.** Subtract: $\begin{array}{r} 14 \\ -\ 3 \\ \hline 11 \end{array}$
4. Subtract: $\begin{array}{r} 54 \\ -21 \\ \hline 33 \end{array}$	**5.** Subtract: $\begin{array}{r} 25 \\ -\ 7 \\ \hline 18 \end{array}$	**6.** Subtract: $\begin{array}{r} 44 \\ -27 \\ \hline 17 \end{array}$
7. Subtract: $\begin{array}{r} 365 \\ -\ 7 \\ \hline 358 \end{array}$	**8.** Subtract: $\begin{array}{r} 325 \\ -\ 81 \\ \hline 244 \end{array}$	**9.** Subtract: $\begin{array}{r} 750 \\ -\ 83 \\ \hline 667 \end{array}$

Pre and Post Assessment for Subtracting Multi-Digit Numbers
ANSWER KEY

Name: ______________________________ Date: ______________

10. Subtract: $\begin{array}{r} 456 \\ -123 \\ \hline 333 \end{array}$	**11.** Subtract: $\begin{array}{r} 651 \\ -114 \\ \hline 537 \end{array}$	**12.** Subtract: $\begin{array}{r} 635 \\ -387 \\ \hline 248 \end{array}$
13. Subtract: $\begin{array}{r} 403 \\ -27 \\ \hline 376 \end{array}$	**14.** Subtract: $\begin{array}{r} 200 \\ -68 \\ \hline 132 \end{array}$	

Common Errors
Subtraction – Whole Numbers

Analysis of student computation errors will often reveal the root of students' misconceptions. When many students in a class are making the same error, the teacher may want to examine the method being used or the explanation given to students for that step. As teachers, what we say is perfectly clear to us, but students may interpret what is said differently.

Very often lack of organization and alignment of place value digits contributes to errors. Large graph paper may help students learn to line up digits and place new, regrouped digits in the correct column.

Shortcuts, that we all develop as adults, may confuse students who are just beginning to learn to regroup. Examine your own methods for subtracting with regrouping and identify where instruction might be confusing to students.

Misconceptions about the concept of subtraction

- No understanding that subtraction means taking away a part
- Difficulty moving between horizontal and vertical format

Misconceptions about modeling

- Difficulty counting backward from a given number
- Difficulty counting backward across a multiple of 10
- Catching and counting backward incorrectly, or using a number line incorrectly resulting in the answer being one too great or one less than the correct difference
- Difficulty aligning digits in a place value column

Misconceptions about regrouping

- Subtracting larger digits from smaller digits no matter which is the minuend, and which is the subtrahend (the minuend is the whole and the subtrahend is the number being subtracted)
- Adding extra value to one place value but not compensating for the regrouping in the adjacent place value to the left
- Compensating for regrouping when necessary but not adding extra value to the necessary digit to the right
- Regrouping but not using the regrouped digits
- Placing regrouped digits in incorrect place value columns
- Regrouping in one place value but not being able to regroup in two place values
- Difficulty regrouping across zeros

Subtraction Skills Trajectory Tracker: Whole Numbers

Student	Level 1 Counting backwards	Level 2 Subtracting single digits	Level 3 Two-digit minus one digit, no regrouping	Level 4 Two-digit minus two digit, no regrouping	Level 5 Two-digit minus one digit, regrouping one's place	Level 6 Two-digit minus two digit, regrouping one's place	Level 7 Three-digit minus one-digit, with and without regrouping one's place	Level 8 Three-digit minus two-digit, regrouping ten's place	Level 9 Three-digit minus two-digit, regrouping two place values	Level 10 Three-digit minus three-digit, no regrouping	Level 11 Three-digit minus three-digit, regrouping one place value	Level 12 Three-digit minus three-digit, regrouping two place values	Level 13 Regrouping across one zero	Level 14 Regrouping across two zeros

Subtraction Trajectory Leveled Practice Sheets

Level 1a Practice Subtracting Multi-Digit Numbers (Counting back from any number 0-100)

Name: ______________________________ Date: ______________

A.	B.	C.
What number comes before 7? ______	What number comes before 13? ______	What number comes before 38? ______
D.	**E.**	**F.**
What number comes before 81? ______	Count backwards: ___, ___, ___, ___, 6	Count backwards: ___, ___, ___, ___, 25
G.	**H.**	**I.**
Count backwards: ___, ___, ___, ___, 30	Count backwards: ___, ___, ___, ___, 67	Count backwards: ___, ___, ___, ___, 73

Level 1b Practice Subtracting Multi-Digit Numbers (Counting back from any number 0-100)

Name: ______________________________ Date: ______________

A. What number comes before 8? ______	**B.** What number comes before 15? ______	**C.** What number comes before 37? ______
D. What number comes before 95? ______	**E.** Count backwards: ___, ___, ___, ___, 7	**F.** Count backwards: ___, ___, ___, ___, 26
G. Count backwards: ___, ___, ___, ___, 40	**H.** Count backwards: ___, ___, ___, ___, 65	**I.** Count backwards: ___, ___, ___, ___, 83

Level 2a Practice Subtracting Multi-Digit Numbers (Subtracting single digits both horizontally and vertically with or without manipulatives)

Name: ______________________________ Date: ______________

A.	B.	C.
Subtract: $3 - 1 =$ ____	Subtract: $4 - 2 =$ ____	Subtract: $5 - 3 =$ ____

D.	E.	F.
Subtract: $7 - 4 =$ ____	Subtract: $\begin{array}{r} 7 \\ -\ 0 \\ \hline \end{array}$	Subtract: $\begin{array}{r} 9 \\ -\ 5 \\ \hline \end{array}$

G.	H.	I.
Subtract: $\begin{array}{r} 8 \\ -\ 4 \\ \hline \end{array}$	Subtract: $\begin{array}{r} 9 \\ -\ 6 \\ \hline \end{array}$	Subtract: $\begin{array}{r} 6 \\ -\ 1 \\ \hline \end{array}$

Level 2b Practice Subtracting Multi-Digit Numbers (Subtracting single digits both horizontally and vertically with or without manipulatives)

Name: ______________________ Date: ______________

A.	B.	C.
Subtract: $2 - 1 =$ ____	Subtract: $5 - 2 =$ ____	Subtract: $4 - 3 =$ ____
D.	**E.**	**F.**
Subtract: $6 - 4 =$ ____	Subtract: $\begin{array}{r} 5 \\ -\ 0 \\ \hline \end{array}$	Subtract: $\begin{array}{r} 8 \\ -\ 3 \\ \hline \end{array}$
G.	**H.**	**I.**
Subtract: $\begin{array}{r} 9 \\ -\ 4 \\ \hline \end{array}$	Subtract: $\begin{array}{r} 8 \\ -\ 6 \\ \hline \end{array}$	Subtract: $\begin{array}{r} 8 \\ -\ 1 \\ \hline \end{array}$

Level 1 & 2 Mixed Practice Subtracting Multi-Digit Numbers (Counting back, Subtracting single digits both horizontally and vertically)

Name: ______________________________ Date: ______________

A.	B.	C.
What number comes before 7? _____	What number comes before 18? _____	What number comes before 80? _____

D.	E.	F.
Count backwards: ___, ___, ___, ___, 9	Subtract: $6 - 0 =$ _____	Subtract: $5 - 3 =$ _____

G.	H.	I.
Subtract: $\begin{array}{r} 8 \\ -\ 6 \\ \hline \end{array}$	Subtract: $\begin{array}{r} 9 \\ -\ 4 \\ \hline \end{array}$	Subtract: $\begin{array}{r} 7 \\ -\ 1 \\ \hline \end{array}$

Level 3a Practice Subtracting Multi-Digit Numbers (Two-digit minus one-digit, no regrouping)

Name: ______________________ Date: ______________

A. Subtract: 15 − 3	**B.** Subtract: 27 − 3	**C.** Subtract: 38 − 3
D. Subtract: 87 − 3	**E.** Subtract: 37 − 7	**F.** Subtract: 45 − 4
G. Subtract: 19 − 5	**H.** Subtract: 66 − 3	**I.** Subtract: 48 − 7

Level 3b Practice Subtracting Multi-Digit Numbers
(Two-digit minus one-digit, no regrouping)

Name: ______________________________ Date: ______________

A.	B.	C.
Subtract: $\begin{array}{r} 14 \\ -\ 3 \\ \hline \end{array}$	Subtract: $\begin{array}{r} 26 \\ -\ 2 \\ \hline \end{array}$	Subtract: $\begin{array}{r} 35 \\ -\ 3 \\ \hline \end{array}$
D.	**E.**	**F.**
Subtract: $\begin{array}{r} 86 \\ -\ 4 \\ \hline \end{array}$	Subtract: $\begin{array}{r} 35 \\ -\ 5 \\ \hline \end{array}$	Subtract: $\begin{array}{r} 49 \\ -\ 8 \\ \hline \end{array}$
G.	**H.**	**I.**
Subtract: $\begin{array}{r} 19 \\ -\ 4 \\ \hline \end{array}$	Subtract: $\begin{array}{r} 63 \\ -\ 2 \\ \hline \end{array}$	Subtract: $\begin{array}{r} 48 \\ -\ 2 \\ \hline \end{array}$

Level 4a Practice Subtracting Multi-Digit Numbers (Two-digit minus two-digit, no regrouping)

Name: ______________________ Date: ______________

A.	B.	C.
Subtract: 45 − 21 ____	Subtract: 53 − 12 ____	Subtract: 34 − 11 ____
D. Subtract: 57 − 30 ____	**E.** Subtract: 54 − 42 ____	**F.** Subtract: 68 − 27 ____
G. Subtract: 78 − 25 ____	**H.** Subtract: 84 − 33 ____	**I.** Subtract: 99 − 17 ____

Level 4b Practice Subtracting Multi-Digit Numbers (Two-digit minus two-digit, no regrouping)

Name: ______________________________ Date: ______________

A.	B.	C.
Subtract: $\begin{array}{r} 36 \\ -21 \\ \hline \end{array}$	Subtract: $\begin{array}{r} 55 \\ -13 \\ \hline \end{array}$	Subtract: $\begin{array}{r} 24 \\ -12 \\ \hline \end{array}$
D.	**E.**	**F.**
Subtract: $\begin{array}{r} 57 \\ -23 \\ \hline \end{array}$	Subtract: $\begin{array}{r} 64 \\ -41 \\ \hline \end{array}$	Subtract: $\begin{array}{r} 68 \\ -40 \\ \hline \end{array}$
G.	**H.**	**I.**
Subtract: $\begin{array}{r} 87 \\ -25 \\ \hline \end{array}$	Subtract: $\begin{array}{r} 74 \\ -22 \\ \hline \end{array}$	Subtract: $\begin{array}{r} 97 \\ -15 \\ \hline \end{array}$

Level 3 & 4 Mixed Practice Subtracting Multi-Digit Numbers (Two-digit minus one– or two-digit, no regrouping)

Name: ______________________ Date: ______________

A.	B.	C.
Subtract: 16 − 3	Subtract: 25 − 2	Subtract: 33 − 3
D.	**E.**	**F.**
Subtract: 39 − 3	Subtract: 64 − 40	Subtract: 68 − 40
G.	**H.**	**I.**
Subtract: 17 − 3	Subtract: 28 − 13	Subtract: 95 − 24

Level 5a Practice Subtracting Multi-Digit Numbers (Two-digit minus one-digit, regrouping one's place)

Name: ______________________________ Date: ________________

A.	B.	C.
Subtract: 45 − 7 ____	Subtract: 32 − 4 ____	Subtract: 23 − 4 ____

D.	E.	F.
Subtract: 32 − 5 ____	Subtract: 64 − 7 ____	Subtract: 68 − 9 ____

G.	H.	I.
Subtract: 41 − 9 ____	Subtract: 40 − 6 ____	Subtract: 90 − 4 ____

Level 5b Practice Subtracting Multi-Digit Numbers (Two-digit minus one-digit, regrouping one's place)

Name: ______________________________ Date: ________________

A.	B.	C.
Subtract: $\begin{array}{r} 35 \\ -\ 6 \\ \hline \end{array}$	Subtract: $\begin{array}{r} 32 \\ -\ 5 \\ \hline \end{array}$	Subtract: $\begin{array}{r} 23 \\ -\ 6 \\ \hline \end{array}$
D.	**E.**	**F.**
Subtract: $\begin{array}{r} 32 \\ -\ 7 \\ \hline \end{array}$	Subtract: $\begin{array}{r} 64 \\ -\ 8 \\ \hline \end{array}$	Subtract: $\begin{array}{r} 61 \\ -\ 3 \\ \hline \end{array}$
G.	**H.**	**I.**
Subtract: $\begin{array}{r} 41 \\ -\ 4 \\ \hline \end{array}$	Subtract: $\begin{array}{r} 20 \\ -\ 3 \\ \hline \end{array}$	Subtract: $\begin{array}{r} 70 \\ -\ 4 \\ \hline \end{array}$

Level 6a Practice Subtracting Multi-Digit Numbers (Two-digit minus two-digit, regrouping one's place)

Name: ______________________________ Date: ________________

A.	B.	C.
Subtract: $\begin{array}{r} 43 \\ -27 \\ \hline \end{array}$	Subtract: $\begin{array}{r} 52 \\ -15 \\ \hline \end{array}$	Subtract: $\begin{array}{r} 26 \\ -17 \\ \hline \end{array}$
D.	**E.**	**F.**
Subtract: $\begin{array}{r} 51 \\ -34 \\ \hline \end{array}$	Subtract: $\begin{array}{r} 34 \\ -25 \\ \hline \end{array}$	Subtract: $\begin{array}{r} 64 \\ -47 \\ \hline \end{array}$
G.	**H.**	**I.**
Subtract: $\begin{array}{r} 46 \\ -17 \\ \hline \end{array}$	Subtract: $\begin{array}{r} 30 \\ -29 \\ \hline \end{array}$	Subtract: $\begin{array}{r} 40 \\ -23 \\ \hline \end{array}$

Level 6b Practice Subtracting Multi-Digit Numbers (Two-digit minus two-digit, regrouping one's place)

Name: ______________________________ Date: ____________________

A. Subtract:	B. Subtract:	C. Subtract:
33 − 25 ——	42 − 16 ——	46 − 27 ——
D. Subtract:	**E.** Subtract:	**F.** Subtract:
71 − 32 ——	34 − 27 ——	64 − 48 ——
G. Subtract:	**H.** Subtract:	**I.** Subtract:
56 − 19 ——	70 − 29 ——	50 − 23 ——

Level 5 & 6 Mixed Practice Subtracting Multi-Digit Numbers (Two-digit minus one– or two-digit, no regrouping)

Name: ______________________________ Date: ____________________

A. Subtract: $\begin{array}{r} 52 \\ -25 \\ \hline \end{array}$	**B.** Subtract: $\begin{array}{r} 42 \\ -\ 6 \\ \hline \end{array}$	**C.** Subtract: $\begin{array}{r} 46 \\ -18 \\ \hline \end{array}$
D. Subtract: $\begin{array}{r} 61 \\ -34 \\ \hline \end{array}$	**E.** Subtract: $\begin{array}{r} 34 \\ -\ 7 \\ \hline \end{array}$	**F.** Subtract: $\begin{array}{r} 64 \\ -\ 8 \\ \hline \end{array}$
G. Subtract: $\begin{array}{r} 51 \\ -14 \\ \hline \end{array}$	**H.** Subtract: $\begin{array}{r} 50 \\ -\ 9 \\ \hline \end{array}$	**I.** Subtract: $\begin{array}{r} 40 \\ -23 \\ \hline \end{array}$

Level 7a Practice Subtracting Multi-Digit Numbers (Three-digit minus one-digit with and without regrouping)

Name: ______________________ Date: ______________

A.	B.	C.
Subtract: $\begin{array}{r} 125 \\ -\ \ 3 \\ \hline \end{array}$	Subtract: $\begin{array}{r} 343 \\ -\ \ 2 \\ \hline \end{array}$	Subtract: $\begin{array}{r} 167 \\ -\ \ 5 \\ \hline \end{array}$
D.	**E.**	**F.**
Subtract: $\begin{array}{r} 328 \\ -\ \ 3 \\ \hline \end{array}$	Subtract: $\begin{array}{r} 655 \\ -\ \ 6 \\ \hline \end{array}$	Subtract: $\begin{array}{r} 514 \\ -\ \ 7 \\ \hline \end{array}$
G.	**H.**	**I.**
Subtract: $\begin{array}{r} 868 \\ -\ \ 9 \\ \hline \end{array}$	Subtract: $\begin{array}{r} 515 \\ -\ \ 7 \\ \hline \end{array}$	Subtract: $\begin{array}{r} 731 \\ -\ \ 0 \\ \hline \end{array}$

Level 7b Practice Subtracting Multi-Digit Numbers (Three-digit minus one-digit with and without regrouping)

Name: ______________________________ Date: ______________

A.	B.	C.
Subtract: $\begin{array}{r} 235 \\ -\quad 2 \\ \hline \end{array}$	Subtract: $\begin{array}{r} 145 \\ -\quad 8 \\ \hline \end{array}$	Subtract: $\begin{array}{r} 568 \\ -\quad 5 \\ \hline \end{array}$
D.	**E.**	**F.**
Subtract: $\begin{array}{r} 367 \\ -\quad 9 \\ \hline \end{array}$	Subtract: $\begin{array}{r} 676 \\ -\quad 6 \\ \hline \end{array}$	Subtract: $\begin{array}{r} 314 \\ -\quad 8 \\ \hline \end{array}$
G.	**H.**	**I.**
Subtract: $\begin{array}{r} 928 \\ -\quad 4 \\ \hline \end{array}$	Subtract: $\begin{array}{r} 437 \\ -\quad 8 \\ \hline \end{array}$	Subtract: $\begin{array}{r} 534 \\ -\quad 0 \\ \hline \end{array}$

Level 8a Practice Subtracting Multi-Digit Numbers (Three-digit minus two-digit, regrouping ten's place)

Name: ______________________________ Date: ________________

A.	B.	C.
Subtract: 746 − 63	Subtract: 253 − 62	Subtract: 726 − 83
D. Subtract: 475 − 83	**E.** Subtract: 204 − 51	**F.** Subtract: 328 − 46
G. Subtract: 623 − 73	**H.** Subtract: 817 − 83	**I.** Subtract: 329 − 46

Level 8b Practice Subtracting Multi-Digit Numbers (Three-digit minus two-digit, regrouping ten's place)

Name: ______________________________ Date: ________________

A. Subtract: 256 − 71	**B.** Subtract: 347 − 86	**C.** Subtract: 316 − 52
D. Subtract: 305 − 83	**E.** Subtract: 586 − 93	**F.** Subtract: 137 − 87
G. Subtract: 416 − 44	**H.** Subtract: 623 − 50	**I.** Subtract: 124 − 43

Level 7 & 8 Mixed Practice Subtracting Multi-Digit Numbers (Three-digit minus one– or two-digit, with and without regrouping one's or ten's place)

Name: ______________________ Date: ______________

A. Subtract: 256 − 2	**B.** Subtract: 542 − 61	**C.** Subtract: 426 − 7
D. Subtract: 624 − 43	**E.** Subtract: 306 − 23	**F.** Subtract: 127 − 9
G. Subtract: 236 − 83	**H.** Subtract: 349 − 90	**I.** Subtract: 220 − 40

Level 9a Practice Subtracting Multi-Digit Numbers (Three-digit minus two-digit, regrouping two place values)

Name: ______________________ Date: ______________

A.	B.	C.
Subtract: $\begin{array}{r} 251 \\ -\ 77 \\ \hline \end{array}$	Subtract: $\begin{array}{r} 342 \\ -\ 89 \\ \hline \end{array}$	Subtract: $\begin{array}{r} 512 \\ -\ 56 \\ \hline \end{array}$
D.	**E.**	**F.**
Subtract: $\begin{array}{r} 332 \\ -\ 83 \\ \hline \end{array}$	Subtract: $\begin{array}{r} 581 \\ -\ 93 \\ \hline \end{array}$	Subtract: $\begin{array}{r} 137 \\ -\ 88 \\ \hline \end{array}$
G.	**H.**	**I.**
Subtract: $\begin{array}{r} 414 \\ -\ 46 \\ \hline \end{array}$	Subtract: $\begin{array}{r} 620 \\ -\ 53 \\ \hline \end{array}$	Subtract: $\begin{array}{r} 123 \\ -\ 84 \\ \hline \end{array}$

Level 9b Practice Subtracting Multi-Digit Numbers (Three-digit minus two-digit, regrouping two place values)

Name: ______________________ Date: ______________

A.	B.	C.
Subtract: 251 − 76	Subtract: 346 − 87	Subtract: 412 − 55
D.	**E.**	**F.**
Subtract: 333 − 84	Subtract: 583 − 96	Subtract: 131 − 87
G.	**H.**	**I.**
Subtract: 712 − 44	Subtract: 610 − 21	Subtract: 121 − 83

Level 10a Practice Subtracting Multi-Digit Numbers (Three-digit minus three-digit, no regrouping)

Name: ______________________ Date: ______________

A.	B.	C.
Subtract: 456 − 123	Subtract: 386 − 153	Subtract: 859 − 426
D.	**E.**	**F.**
Subtract: 654 − 231	Subtract: 257 − 213	Subtract: 756 − 454
G.	**H.**	**I.**
Subtract: 492 − 221	Subtract: 656 − 110	Subtract: 933 − 400

Level 10b Practice Subtracting Multi-Digit Numbers (Three-digit minus three-digit, no regrouping)

Name: ______________________________ Date: ____________________

A. Subtract: 456 − 213	**B.** Subtract: 386 − 235	**C.** Subtract: 859 − 241
D. Subtract: 654 − 212	**E.** Subtract: 257 − 123	**F.** Subtract: 756 − 522
G. Subtract: 492 − 112	**H.** Subtract: 656 − 230	**I.** Subtract: 945 − 300

Level 9 & 10 Mixed Practice Subtracting Multi-Digit Numbers (Three-digit minus two or three-digit, with and without regrouping)

Name: ______________________________ Date: ____________________

A.	B.	C.
Subtract: 456 − 77 ____	Subtract: 572 − 341 ____	Subtract: 811 − 49 ____
D. Subtract: 754 − 265 ____	**E.** Subtract: 311 − 246 ____	**F.** Subtract: 632 − 86 ____
G. Subtract: 912 − 83 ____	**H.** Subtract: 286 − 140 ____	**I.** Subtract: 515 − 200 ____

Level 11a Practice Subtracting Multi-Digit Numbers (Three-digit minus three-digit, regrouping one place value)

Name: ______________________________ Date: ________________

A.	B.	C.
Subtract: 456 − 217	Subtract: 381 − 233	Subtract: 352 − 214
D. Subtract: 414 − 121	**E.** Subtract: 826 − 236	**F.** Subtract: 656 − 238
G. Subtract: 428 − 137	**H.** Subtract: 608 − 156	**I.** Subtract: 920 − 317

Level 11b Practice Subtracting Multi-Digit Numbers (Three-digit minus three-digit, regrouping one place value)

Name: ______________________________ Date: ____________________

A.	B.	C.

Subtract:
 	
  358
-277
	Subtract:
	
	  432
	-214
		Subtract:
		
		  717
		-362

D.
Subtract:
 	
  414
-152
	E.
	Subtract:
	
	  635
	-264
		F.
		Subtract:
		
		  544
		-438

G.
Subtract:
 	
  418
-137
	H.
	Subtract:
	
	  880
	-234
		I.
		Subtract:
		
		  940
		-333

Level 12a Practice Subtracting Multi-Digit Numbers (Three-digit minus three-digit, regrouping two place values)

Name: ______________________ Date: ______________

A.	B.	C.
Subtract: 451 − 184	Subtract: 373 − 296	Subtract: 524 − 278
D.	**E.**	**F.**
Subtract: 621 − 187	Subtract: 532 − 234	Subtract: 967 − 188
G.	**H.**	**I.**
Subtract: 377 − 178	Subtract: 820 − 249	Subtract: 930 − 351

Level 12b Practice Subtracting Multi-Digit Numbers (Three-digit minus three-digit, regrouping two place values)

Name: ______________________________ Date: ____________________

A.	B.	C.
Subtract: $\begin{array}{r} 351 \\ -\,183 \\ \hline \end{array}$	Subtract: $\begin{array}{r} 453 \\ -\,296 \\ \hline \end{array}$	Subtract: $\begin{array}{r} 656 \\ -\,278 \\ \hline \end{array}$
D.	**E.**	**F.**
Subtract: $\begin{array}{r} 463 \\ -\,187 \\ \hline \end{array}$	Subtract: $\begin{array}{r} 612 \\ -\,234 \\ \hline \end{array}$	Subtract: $\begin{array}{r} 673 \\ -\,188 \\ \hline \end{array}$
G.	**H.**	**I.**
Subtract: $\begin{array}{r} 352 \\ -\,174 \\ \hline \end{array}$	Subtract: $\begin{array}{r} 730 \\ -\,159 \\ \hline \end{array}$	Subtract: $\begin{array}{r} 940 \\ -\,362 \\ \hline \end{array}$

Level 11 & 12 Mixed Practice Subtracting Multi-Digit Numbers (Three-digit minus three-digit, regrouping one or two place values)

Name: ______________________ Date: ______________

A. Subtract: $\begin{array}{r} 491 \\ -383 \\ \hline \end{array}$	**B.** Subtract: $\begin{array}{r} 715 \\ -234 \\ \hline \end{array}$	**C.** Subtract: $\begin{array}{r} 958 \\ -267 \\ \hline \end{array}$
D. Subtract: $\begin{array}{r} 812 \\ -224 \\ \hline \end{array}$	**E.** Subtract: $\begin{array}{r} 473 \\ -135 \\ \hline \end{array}$	**F.** Subtract: $\begin{array}{r} 338 \\ -184 \\ \hline \end{array}$
G. Subtract: $\begin{array}{r} 951 \\ -175 \\ \hline \end{array}$	**H.** Subtract: $\begin{array}{r} 750 \\ -223 \\ \hline \end{array}$	**I.** Subtract: $\begin{array}{r} 423 \\ -184 \\ \hline \end{array}$

Level 13a Practice Subtracting Multi-Digit Numbers (Regrouping across one zero)

Name: ______________________ Date: ______________

A.	B.	C.
Subtract: $\begin{array}{r} 401 \\ -184 \\ \hline \end{array}$	Subtract: $\begin{array}{r} 802 \\ -226 \\ \hline \end{array}$	Subtract: $\begin{array}{r} 504 \\ -278 \\ \hline \end{array}$
D.	**E.**	**F.**
Subtract: $\begin{array}{r} 601 \\ -187 \\ \hline \end{array}$	Subtract: $\begin{array}{r} 503 \\ -234 \\ \hline \end{array}$	Subtract: $\begin{array}{r} 905 \\ -188 \\ \hline \end{array}$
G.	**H.**	**I.**
Subtract: $\begin{array}{r} 307 \\ -178 \\ \hline \end{array}$	Subtract: $\begin{array}{r} 808 \\ -249 \\ \hline \end{array}$	Subtract: $\begin{array}{r} 904 \\ -356 \\ \hline \end{array}$

Level 13b Practice Subtracting Multi-Digit Numbers (Regrouping across one zero)

Name: ______________________________ Date: ____________________

A. Subtract: 304 – 126	**B.** Subtract: 407 – 216	**C.** Subtract: 402 – 258
D. Subtract: 405 – 187	**E.** Subtract: 302 – 214	**F.** Subtract: 905 – 177
G. Subtract: 302 – 176	**H.** Subtract: 803 – 217	**I.** Subtract: 901 – 359

Level 14a Practice Subtracting Multi-Digit Numbers (Regrouping across two zeros)

Name: ______________________________ Date: ____________________

A.	B.	C.
Subtract: $\begin{array}{r} 400 \\ -184 \\ \hline \end{array}$	Subtract: $\begin{array}{r} 800 \\ -226 \\ \hline \end{array}$	Subtract: $\begin{array}{r} 500 \\ -278 \\ \hline \end{array}$
D.	**E.**	**F.**
Subtract: $\begin{array}{r} 600 \\ -187 \\ \hline \end{array}$	Subtract: $\begin{array}{r} 500 \\ -234 \\ \hline \end{array}$	Subtract: $\begin{array}{r} 900 \\ -188 \\ \hline \end{array}$
G.	**H.**	**I.**
Subtract: $\begin{array}{r} 300 \\ -178 \\ \hline \end{array}$	Subtract: $\begin{array}{r} 800 \\ -249 \\ \hline \end{array}$	Subtract: $\begin{array}{r} 900 \\ -351 \\ \hline \end{array}$

Level 14b Practice Subtracting Multi-Digit Numbers (Regrouping across two zeros)

Name: ______________________________ Date: ________________

A.	B.	C.
Subtract: $\begin{array}{r} 300 \\ -126 \\ \hline \end{array}$	Subtract: $\begin{array}{r} 400 \\ -216 \\ \hline \end{array}$	Subtract: $\begin{array}{r} 400 \\ -258 \\ \hline \end{array}$
D.	**E.**	**F.**
Subtract: $\begin{array}{r} 500 \\ -187 \\ \hline \end{array}$	Subtract: $\begin{array}{r} 300 \\ -214 \\ \hline \end{array}$	Subtract: $\begin{array}{r} 900 \\ -177 \\ \hline \end{array}$
G.	**H.**	**I.**
Subtract: $\begin{array}{r} 300 \\ -176 \\ \hline \end{array}$	Subtract: $\begin{array}{r} 800 \\ -211 \\ \hline \end{array}$	Subtract: $\begin{array}{r} 900 \\ -359 \\ \hline \end{array}$

Level 13 & 14 Mixed Practice Subtracting Multi-Digit Numbers (Regrouping across one and two zeros)

Name: ______________________________ Date: ________________

A.	B.	C.
Subtract: $\begin{array}{r} 401 \\ -383 \\ \hline \end{array}$	Subtract: $\begin{array}{r} 705 \\ -216 \\ \hline \end{array}$	Subtract: $\begin{array}{r} 900 \\ -267 \\ \hline \end{array}$
D.	**E.**	**F.**
Subtract: $\begin{array}{r} 800 \\ -224 \\ \hline \end{array}$	Subtract: $\begin{array}{r} 403 \\ -115 \\ \hline \end{array}$	Subtract: $\begin{array}{r} 302 \\ -184 \\ \hline \end{array}$
G.	**H.**	**I.**
Subtract: $\begin{array}{r} 901 \\ -175 \\ \hline \end{array}$	Subtract: $\begin{array}{r} 700 \\ -223 \\ \hline \end{array}$	Subtract: $\begin{array}{r} 400 \\ -184 \\ \hline \end{array}$

Subtraction Trajectory Answer Keys

Level 1a Practice Subtracting Multi-Digit Numbers (Counting back from any number 0-100)
ANSWER KEY

Name: ______________________ Date: ______________

A.	B.	C.
What number comes before 7? 6	What number comes before 13? 12	What number comes before 38? 37
D.	**E.**	**F.**
What number comes before 81? 80	Count backwards: 2, 3, 4, 5, 6	Count backwards: 21, 22, 23, 24, 25
G.	**H.**	**I.**
Count backwards: 26, 27, 28, 29, 30	Count backwards: 63, 64, 65, 66, 67	Count backwards: 69, 70, 71, 72, 73

Level 1b Practice Subtracting Multi-Digit Numbers (Counting back from any number 0-100) ANSWER KEY

Name: ______________________ Date: ______________

A.	B.	C.
What number comes before 8? ___7___	What number comes before 15? ___14___	What number comes before 37? ___36___
D.	**E.**	**F.**
What number comes before 95? ___94___	Count backwards: _3_, _4_, _5_, _6_, 7	Count backwards: _22_, _23_, _24_, _25_, 26
G.	**H.**	**I.**
Count backwards: _36_, _37_, _38_, _39_, 40	Count backwards: _61_, _62_, _63_, _64_, 65	Count backwards: _79_, _80_, _81_, _82_, 83

Level 2a Practice Subtracting Multi-Digit Numbers (Subtracting single digits both horizontally and vertically with or without manipulatives)
ANSWER KEY

Name: ______________________ Date: ______________

A.	B.	C.
Subtract: $3 - 1 = \underline{2}$	Subtract: $4 - 2 = \underline{2}$	Subtract: $5 - 3 = \underline{2}$

D.	E.	F.
Subtract: $7 - 4 = \underline{3}$	Subtract: $\begin{array}{r} 7 \\ -\ 0 \\ \hline 7 \end{array}$	Subtract: $\begin{array}{r} 9 \\ -\ 5 \\ \hline 4 \end{array}$

G.	H.	I.
Subtract: $\begin{array}{r} 8 \\ -\ 4 \\ \hline 4 \end{array}$	Subtract: $\begin{array}{r} 9 \\ -\ 6 \\ \hline 3 \end{array}$	Subtract: $\begin{array}{r} 6 \\ -\ 1 \\ \hline 5 \end{array}$

Level 2b Practice Subtracting Multi-Digit Numbers (Subtracting single digits both horizontally and vertically with or without manipulatives) ANSWER KEY

Name: ______________________________ Date: ____________________

A. Subtract: 2 – 1 = <u>1</u>	**B.** Subtract: 5 – 2 = <u>3</u>	**C.** Subtract: 4 – 3 = <u>1</u>
D. Subtract: 6 – 4 = <u>2</u>	**E.** Subtract: $\begin{array}{r} 5 \\ -\ 0 \\ \hline 5 \end{array}$	**F.** Subtract: $\begin{array}{r} 8 \\ -\ 3 \\ \hline 5 \end{array}$
G. Subtract: $\begin{array}{r} 9 \\ -\ 4 \\ \hline 5 \end{array}$	**H.** Subtract: $\begin{array}{r} 8 \\ -\ 6 \\ \hline 2 \end{array}$	**I.** Subtract: $\begin{array}{r} 8 \\ -\ 1 \\ \hline 7 \end{array}$

Level 1 & 2 Mixed Practice Subtracting Multi-Digit Numbers (Counting back, Subtracting single digits both horizontally and vertically) ANSWER KEY

Name: ______________________________ Date: ______________

A. What number comes before 7? _6_	**B.** What number comes before 18? _17_	**C.** What number comes before 80? _79_
D. Count backwards: _5_, _6_, _7_, _8_, 9	**E.** Subtract: 6 – 0 = _6_	**F.** Subtract: 5 – 3 = _2_
G. Subtract: $\begin{array}{r} 8 \\ -6 \\ \hline 2 \end{array}$	**H.** Subtract: $\begin{array}{r} 9 \\ -4 \\ \hline 5 \end{array}$	**I.** Subtract: $\begin{array}{r} 7 \\ -1 \\ \hline 6 \end{array}$

Level 3a Practice Subtracting Multi-Digit Numbers
(Two-digit minus one-digit, no regrouping)
ANSWER KEY

Name: ____________________ Date: ____________________

A. Subtract: 15 − 3 12	**B.** Subtract: 27 − 3 24	**C.** Subtract: 38 − 3 35
D. Subtract: 87 − 3 84	**E.** Subtract: 37 − 7 30	**F.** Subtract: 45 − 4 41
G. Subtract: 19 − 5 14	**H.** Subtract: 66 − 3 63	**I.** Subtract: 48 − 7 41

Level 3b Practice Subtracting Multi-Digit Numbers
(Two-digit minus one-digit, no regrouping)
ANSWER KEY

Name: ______________________________ Date: ________________

A. Subtract: 14 − 3 11	**B.** Subtract: 26 − 2 24	**C.** Subtract: 35 − 3 32
D. Subtract: 86 − 4 82	**E.** Subtract: 35 − 5 30	**F.** Subtract: 49 − 8 41
G. Subtract: 19 − 4 15	**H.** Subtract: 63 − 2 61	**I.** Subtract: 48 − 2 46

Level 4a Practice Subtracting Multi-Digit Numbers (Two-digit minus two-digit, no regrouping) ANSWER KEY

Name: ______________________________ Date: ________________

A.

Subtract:

$$\begin{array}{r} 45 \\ -\,21 \\ \hline 24 \end{array}$$

B.

Subtract:

$$\begin{array}{r} 53 \\ -\,12 \\ \hline 41 \end{array}$$

C.

Subtract:

$$\begin{array}{r} 34 \\ -\,11 \\ \hline 23 \end{array}$$

D.

Subtract:

$$\begin{array}{r} 57 \\ -\,30 \\ \hline 27 \end{array}$$

E.

Subtract:

$$\begin{array}{r} 54 \\ -\,42 \\ \hline 12 \end{array}$$

F.

Subtract:

$$\begin{array}{r} 68 \\ -\,27 \\ \hline 41 \end{array}$$

G.

Subtract:

$$\begin{array}{r} 78 \\ -\,25 \\ \hline 53 \end{array}$$

H.

Subtract:

$$\begin{array}{r} 84 \\ -\,33 \\ \hline 51 \end{array}$$

I.

Subtract:

$$\begin{array}{r} 99 \\ -\,17 \\ \hline 82 \end{array}$$

Level 4b Practice Subtracting Multi-Digit Numbers (Two-digit minus two-digit, no regrouping) ANSWER KEY

Name: ______________________ Date: ______________

A. Subtract: $\begin{array}{r} 36 \\ -21 \\ \hline 15 \end{array}$	**B.** Subtract: $\begin{array}{r} 55 \\ -13 \\ \hline 42 \end{array}$	**C.** Subtract: $\begin{array}{r} 24 \\ -12 \\ \hline 12 \end{array}$
D. Subtract: $\begin{array}{r} 57 \\ -23 \\ \hline 34 \end{array}$	**E.** Subtract: $\begin{array}{r} 64 \\ -41 \\ \hline 23 \end{array}$	**F.** Subtract: $\begin{array}{r} 68 \\ -40 \\ \hline 28 \end{array}$
G. Subtract: $\begin{array}{r} 87 \\ -25 \\ \hline 62 \end{array}$	**H.** Subtract: $\begin{array}{r} 74 \\ -22 \\ \hline 52 \end{array}$	**I.** Subtract: $\begin{array}{r} 97 \\ -15 \\ \hline 82 \end{array}$

Level 3 & 4 Mixed Practice Subtracting Multi-Digit Numbers (Two-digit minus one– or two-digit, no regrouping) ANSWER KEY

Name: ______________________ Date: ______________

A. Subtract: $\begin{array}{r} 16 \\ -\ 3 \\ \hline 13 \end{array}$	**B.** Subtract: $\begin{array}{r} 25 \\ -\ 2 \\ \hline 23 \end{array}$	**C.** Subtract: $\begin{array}{r} 33 \\ -\ 3 \\ \hline 30 \end{array}$
D. Subtract: $\begin{array}{r} 39 \\ -\ 3 \\ \hline 36 \end{array}$	**E.** Subtract: $\begin{array}{r} 64 \\ -40 \\ \hline 24 \end{array}$	**F.** Subtract: $\begin{array}{r} 68 \\ -40 \\ \hline 28 \end{array}$
G. Subtract: $\begin{array}{r} 17 \\ -\ 3 \\ \hline 14 \end{array}$	**H.** Subtract: $\begin{array}{r} 28 \\ -13 \\ \hline 15 \end{array}$	**I.** Subtract: $\begin{array}{r} 95 \\ -24 \\ \hline 71 \end{array}$

Level 5a Practice Subtracting Multi-Digit Numbers (Two-digit minus one-digit, regrouping one's place)
ANSWER KEY

Name: ______________________________ Date: ________________

A.	B.	C.
Subtract: $\begin{array}{r} 45 \\ -\ 7 \\ \hline 38 \end{array}$	Subtract: $\begin{array}{r} 32 \\ -\ 4 \\ \hline 28 \end{array}$	Subtract: $\begin{array}{r} 23 \\ -\ 4 \\ \hline 19 \end{array}$
D. Subtract: $\begin{array}{r} 32 \\ -\ 5 \\ \hline 27 \end{array}$	**E.** Subtract: $\begin{array}{r} 64 \\ -\ 7 \\ \hline 57 \end{array}$	**F.** Subtract: $\begin{array}{r} 68 \\ -\ 9 \\ \hline 59 \end{array}$
G. Subtract: $\begin{array}{r} 41 \\ -\ 9 \\ \hline 32 \end{array}$	**H.** Subtract: $\begin{array}{r} 40 \\ -\ 6 \\ \hline 34 \end{array}$	**I.** Subtract: $\begin{array}{r} 90 \\ -\ 4 \\ \hline 86 \end{array}$

Level 5b Practice Subtracting Multi-Digit Numbers (Two-digit minus one-digit, regrouping one's place) ANSWER KEY

Name: ______________________________ Date: ________________

A. Subtract: 35 – 6 29	**B.** Subtract: 32 – 5 27	**C.** Subtract: 23 – 6 17
D. Subtract: 32 – 7 25	**E.** Subtract: 64 – 8 56	**F.** Subtract: 61 – 3 58
G. Subtract: 41 – 4 37	**H.** Subtract: 20 – 3 17	**I.** Subtract: 70 – 4 66

Level 6a Practice Subtracting Multi-Digit Numbers (Two-digit minus two-digit, regrouping one's place) ANSWER KEY

Name: ______________________ Date: ______________

A.	B.	C.
Subtract: $\begin{array}{r} 43 \\ -27 \\ \hline 16 \end{array}$	Subtract: $\begin{array}{r} 52 \\ -15 \\ \hline 37 \end{array}$	Subtract: $\begin{array}{r} 26 \\ -17 \\ \hline 9 \end{array}$
D.	**E.**	**F.**
Subtract: $\begin{array}{r} 51 \\ -34 \\ \hline 17 \end{array}$	Subtract: $\begin{array}{r} 34 \\ -25 \\ \hline 9 \end{array}$	Subtract: $\begin{array}{r} 64 \\ -47 \\ \hline 17 \end{array}$
G.	**H.**	**I.**
Subtract: $\begin{array}{r} 46 \\ -17 \\ \hline 29 \end{array}$	Subtract: $\begin{array}{r} 30 \\ -29 \\ \hline 1 \end{array}$	Subtract: $\begin{array}{r} 40 \\ -23 \\ \hline 17 \end{array}$

Level 6b Practice Subtracting Multi-Digit Numbers (Two-digit minus two-digit, regrouping one's place) ANSWER KEY

Name: ______________________ Date: ______________

A. Subtract: 33 − 25 8	**B.** Subtract: 42 − 16 26	**C.** Subtract: 46 − 27 19
D. Subtract: 71 − 32 39	**E.** Subtract: 34 − 27 7	**F.** Subtract: 64 − 48 16
G. Subtract: 56 − 19 37	**H.** Subtract: 70 − 29 41	**I.** Subtract: 50 − 23 27

Level 5 & 6 Mixed Practice Subtracting Multi-Digit Numbers (Two-digit minus one– or two-digit, regrouping one's place)
ANSWER KEY

Name: ______________________________ Date: ______________

A. Subtract: $\begin{array}{r} 52 \\ -25 \\ \hline 27 \end{array}$	**B.** Subtract: $\begin{array}{r} 42 \\ -\ 6 \\ \hline 36 \end{array}$	**C.** Subtract: $\begin{array}{r} 46 \\ -18 \\ \hline 28 \end{array}$
D. Subtract: $\begin{array}{r} 61 \\ -34 \\ \hline 27 \end{array}$	**E.** Subtract: $\begin{array}{r} 34 \\ -\ 7 \\ \hline 27 \end{array}$	**F.** Subtract: $\begin{array}{r} 64 \\ -\ 8 \\ \hline 56 \end{array}$
G. Subtract: $\begin{array}{r} 51 \\ -14 \\ \hline 37 \end{array}$	**H.** Subtract: $\begin{array}{r} 50 \\ -\ 9 \\ \hline 41 \end{array}$	**I.** Subtract: $\begin{array}{r} 40 \\ -23 \\ \hline 17 \end{array}$

Level 7a Practice Subtracting Multi-Digit Numbers (Three-digit minus one-digit with and without regrouping) ANSWER KEY

Name: ______________________________ Date: ______________

A.	B.	C.
Subtract: $\begin{array}{r} 125 \\ -\quad 3 \\ \hline 122 \end{array}$	Subtract: $\begin{array}{r} 343 \\ -\quad 2 \\ \hline 341 \end{array}$	Subtract: $\begin{array}{r} 167 \\ -\quad 5 \\ \hline 162 \end{array}$
D.	**E.**	**F.**
Subtract: $\begin{array}{r} 328 \\ -\quad 3 \\ \hline 325 \end{array}$	Subtract: $\begin{array}{r} 655 \\ -\quad 6 \\ \hline 649 \end{array}$	Subtract: $\begin{array}{r} 514 \\ -\quad 7 \\ \hline 507 \end{array}$
G.	**H.**	**I.**
Subtract: $\begin{array}{r} 868 \\ -\quad 9 \\ \hline 859 \end{array}$	Subtract: $\begin{array}{r} 515 \\ -\quad 7 \\ \hline 508 \end{array}$	Subtract: $\begin{array}{r} 731 \\ -\quad 0 \\ \hline 731 \end{array}$

Level 7b Practice Subtracting Multi-Digit Numbers (Three-digit minus one-digit with and without regrouping) ANSWER KEY

Name: ______________________ Date: ______________

A.	B.	C.
Subtract: $\begin{array}{r} 235 \\ -\ \ 2 \\ \hline 233 \end{array}$	Subtract: $\begin{array}{r} 145 \\ -\ \ 8 \\ \hline 137 \end{array}$	Subtract: $\begin{array}{r} 568 \\ -\ \ 5 \\ \hline 563 \end{array}$
D.	**E.**	**F.**
Subtract: $\begin{array}{r} 367 \\ -\ \ 9 \\ \hline 358 \end{array}$	Subtract: $\begin{array}{r} 676 \\ -\ \ 6 \\ \hline 670 \end{array}$	Subtract: $\begin{array}{r} 314 \\ -\ \ 8 \\ \hline 306 \end{array}$
G.	**H.**	**I.**
Subtract: $\begin{array}{r} 928 \\ -\ \ 4 \\ \hline 924 \end{array}$	Subtract: $\begin{array}{r} 437 \\ -\ \ 8 \\ \hline 429 \end{array}$	Subtract: $\begin{array}{r} 534 \\ -\ \ 0 \\ \hline 534 \end{array}$

Level 8a Practice Subtracting Multi-Digit Numbers (Three-digit minus two-digit, regrouping ten's place) ANSWER KEY

Name: ______________________ Date: ______________

A. Subtract: 746 – 63 683	**B.** Subtract: 253 – 62 191	**C.** Subtract: 726 – 83 643
D. Subtract: 475 – 83 392	**E.** Subtract: 204 – 51 153	**F.** Subtract: 328 – 46 282
G. Subtract: 623 – 73 550	**H.** Subtract: 817 – 83 734	**I.** Subtract: 329 – 46 283

Level 8b Practice Subtracting Multi-Digit Numbers (Three-digit minus two-digit, regrouping ten's place) ANSWER KEY

Name: ______________________________ Date: ______________

A.	B.	C.
Subtract: 256 − 71 = 185	Subtract: 347 − 86 = 261	Subtract: 316 − 52 = 264

D.	E.	F.
Subtract: 305 − 83 = 222	Subtract: 586 − 93 = 493	Subtract: 137 − 87 = 50

G.	H.	I.
Subtract: 416 − 44 = 372	Subtract: 623 − 50 = 573	Subtract: 124 − 43 = 81

Level 7 & 8 Mixed Practice Subtracting Multi-Digit Numbers (Three-digit minus one– or two-digit, with and without regrouping one's or ten's place) ANSWER KEY

Name: ______________________ Date: ______________

A.	B.	C.
Subtract: 256 − 2 = 254	Subtract: 542 − 61 = 481	Subtract: 426 − 7 = 419

D.	E.	F.
Subtract: 624 − 43 = 581	Subtract: 306 − 23 = 283	Subtract: 127 − 9 = 118

G.	H.	I.
Subtract: 236 − 83 = 153	Subtract: 349 − 90 = 259	Subtract: 220 − 40 = 180

Level 9a Practice Subtracting Multi-Digit Numbers
(Three-digit minus two-digit, regrouping two place values)
ANSWER KEY

Name: ______________________________ Date: ______________

A.	B.	C.
Subtract: 251 − 77 = 174	Subtract: 342 − 89 = 253	Subtract: 512 − 56 = 456

D.	E.	F.
Subtract: 332 − 83 = 249	Subtract: 581 − 93 = 488	Subtract: 137 − 88 = 49

G.	H.	I.
Subtract: 414 − 46 = 368	Subtract: 620 − 53 = 567	Subtract: 123 − 84 = 39

Level 9b Practice Subtracting Multi-Digit Numbers (Three-digit minus two-digit, regrouping two place values) ANSWER KEY

Name: ____________________ Date: ____________

A.	B.	C.
Subtract: 251 − 76 = 175	Subtract: 346 − 87 = 259	Subtract: 412 − 55 = 357
D.	**E.**	**F.**
Subtract: 333 − 84 = 249	Subtract: 583 − 96 = 487	Subtract: 131 − 87 = 44
G.	**H.**	**I.**
Subtract: 712 − 44 = 668	Subtract: 610 − 21 = 589	Subtract: 121 − 83 = 38

Level 10a Practice Subtracting Multi-Digit Numbers (Three-digit minus three-digit, no regrouping) ANSWER KEY

Name: ______________________ Date: ______________

A.	B.	C.
Subtract: 456 − 123 = 333	Subtract: 386 − 153 = 233	Subtract: 859 − 426 = 433

D.	E.	F.
Subtract: 654 − 231 = 423	Subtract: 257 − 213 = 44	Subtract: 756 − 454 = 302

G.	H.	I.
Subtract: 492 − 221 = 271	Subtract: 656 − 110 = 546	Subtract: 933 − 400 = 533

Level 10b Practice Subtracting Multi-Digit Numbers (Three-digit minus three-digit, no regrouping) ANSWER KEY

Name: ______________________ Date: ______________

A.	B.	C.
Subtract: 456 − 213 = 243	Subtract: 386 − 235 = 151	Subtract: 859 − 241 = 618
D.	**E.**	**F.**
Subtract: 654 − 212 = 442	Subtract: 257 − 123 = 134	Subtract: 756 − 522 = 234
G.	**H.**	**I.**
Subtract: 492 − 112 = 380	Subtract: 656 − 230 = 426	Subtract: 945 − 300 = 645

Level 9 & 10 Mixed Practice Subtracting Multi-Digit Numbers (Three-digit minus two or three-digit, with and without regrouping) ANSWER KEY

Name: ____________________ Date: ____________

A.	B.	C.
Subtract: 456 − 77 379	Subtract: 572 − 341 231	Subtract: 811 − 49 762
D.	**E.**	**F.**
Subtract: 754 − 265 489	Subtract: 311 − 246 65	Subtract: 632 − 86 546
G.	**H.**	**I.**
Subtract: 912 − 83 829	Subtract: 286 − 140 146	Subtract: 515 − 200 315

Level 11a Practice Subtracting Multi-Digit Numbers (Three-digit minus three-digit, regrouping one place value) ANSWER KEY

Name: ______________________________ Date: ____________________

A. Subtract: 456 – 217 ——— 239	**B.** Subtract: 381 – 233 ——— 148	**C.** Subtract: 352 – 214 ——— 138
D. Subtract: 414 – 121 ——— 293	**E.** Subtract: 826 – 236 ——— 590	**F.** Subtract: 656 – 238 ——— 418
G. Subtract: 428 – 137 ——— 291	**H.** Subtract: 608 – 156 ——— 452	**I.** Subtract: 920 – 317 ——— 603

Level 11b Practice Subtracting Multi-Digit Numbers (Three-digit minus three-digit, regrouping one place value)
ANSWER KEY

Name: ____________________ Date: ____________

A.	B.	C.
Subtract: 358 - 277 81	Subtract: 432 - 214 218	Subtract: 717 - 362 355
D.	**E.**	**F.**
Subtract: 414 - 152 262	Subtract: 635 - 264 371	Subtract: 544 - 438 106
G.	**H.**	**I.**
Subtract: 418 - 137 281	Subtract: 880 - 234 646	Subtract: 940 - 333 607

Level 12a Practice Subtracting Multi-Digit Numbers (Three-digit minus three-digit, regrouping two place values) ANSWER KEY

Name: ______________________ Date: ______________

A.	B.	C.
Subtract: $\begin{array}{r} 451 \\ -184 \\ \hline 267 \end{array}$	Subtract: $\begin{array}{r} 373 \\ -296 \\ \hline 77 \end{array}$	Subtract: $\begin{array}{r} 524 \\ -278 \\ \hline 246 \end{array}$
D.	**E.**	**F.**
Subtract: $\begin{array}{r} 621 \\ -187 \\ \hline 434 \end{array}$	Subtract: $\begin{array}{r} 532 \\ -234 \\ \hline 298 \end{array}$	Subtract: $\begin{array}{r} 967 \\ -188 \\ \hline 779 \end{array}$
G.	**H.**	**I.**
Subtract: $\begin{array}{r} 377 \\ -178 \\ \hline 199 \end{array}$	Subtract: $\begin{array}{r} 820 \\ -249 \\ \hline 571 \end{array}$	Subtract: $\begin{array}{r} 930 \\ -351 \\ \hline 579 \end{array}$

Level 12b Practice Subtracting Multi-Digit Numbers
(Three-digit minus three-digit, regrouping two place values)
ANSWER KEY

Name: ______________________________ Date: ______________

A.	B.	C.
Subtract: $\begin{array}{r} 351 \\ -183 \\ \hline 168 \end{array}$	Subtract: $\begin{array}{r} 453 \\ -296 \\ \hline 157 \end{array}$	Subtract: $\begin{array}{r} 656 \\ -278 \\ \hline 378 \end{array}$
D.	**E.**	**F.**
Subtract: $\begin{array}{r} 463 \\ -187 \\ \hline 276 \end{array}$	Subtract: $\begin{array}{r} 612 \\ -234 \\ \hline 378 \end{array}$	Subtract: $\begin{array}{r} 673 \\ -188 \\ \hline 485 \end{array}$
G.	**H.**	**I.**
Subtract: $\begin{array}{r} 352 \\ -174 \\ \hline 178 \end{array}$	Subtract: $\begin{array}{r} 730 \\ -159 \\ \hline 571 \end{array}$	Subtract: $\begin{array}{r} 940 \\ -362 \\ \hline 578 \end{array}$

Level 11 & 12 Mixed Practice Subtracting Multi-Digit Numbers (Three-digit minus three-digit, regrouping one or two place values) ANSWER KEY

Name: ______________________ Date: ______________

A.	B.	C.
Subtract: $\begin{array}{r} 491 \\ -383 \\ \hline 108 \end{array}$	Subtract: $\begin{array}{r} 715 \\ -234 \\ \hline 481 \end{array}$	Subtract: $\begin{array}{r} 958 \\ -267 \\ \hline 691 \end{array}$
D.	**E.**	**F.**
Subtract: $\begin{array}{r} 812 \\ -224 \\ \hline 588 \end{array}$	Subtract: $\begin{array}{r} 473 \\ -135 \\ \hline 338 \end{array}$	Subtract: $\begin{array}{r} 338 \\ -184 \\ \hline 154 \end{array}$
G.	**H.**	**I.**
Subtract: $\begin{array}{r} 951 \\ -175 \\ \hline 776 \end{array}$	Subtract: $\begin{array}{r} 750 \\ -223 \\ \hline 527 \end{array}$	Subtract: $\begin{array}{r} 423 \\ -184 \\ \hline 239 \end{array}$

Level 13a Practice Subtracting Multi-Digit Numbers (Regrouping across one zero) ANSWER KEY

Name: ______________________ Date: ______________

A.	B.	C.
Subtract: 401 − 184 217	Subtract: 802 − 226 576	Subtract: 504 − 278 226
D. Subtract: 601 − 187 414	**E.** Subtract: 503 − 234 269	**F.** Subtract: 905 − 188 717
G. Subtract: 307 − 178 129	**H.** Subtract: 808 − 249 559	**I.** Subtract: 904 − 356 548

Level 13b Practice Subtracting Multi-Digit Number (Regrouping across one zero) ANSWER KEY

Name: ______________________________ Date: ____________________

A.	B.	C.
Subtract: 304 − 126 178	Subtract: 407 − 216 191	Subtract: 402 − 258 144
D.	**E.**	**F.**
Subtract: 405 − 187 218	Subtract: 302 − 214 88	Subtract: 905 − 177 728
G.	**H.**	**I.**
Subtract: 302 − 176 126	Subtract: 803 − 217 586	Subtract: 901 − 359 542

Level 14a Practice Subtracting Multi-Digit Numbers (Regrouping across two zeros) ANSWER KEY

Name: ______________________ Date: ______________

A. Subtract: 400 − 184 = 216	**B.** Subtract: 800 − 226 = 574	**C.** Subtract: 500 − 278 = 222
D. Subtract: 600 − 187 = 413	**E.** Subtract: 500 − 234 = 266	**F.** Subtract: 900 − 188 = 712
G. Subtract: 300 − 178 = 122	**H.** Subtract: 800 − 249 = 551	**I.** Subtract: 900 − 351 = 549

Level 14b Practice Subtracting Multi-Digit Numbers (Regrouping across two zeros)
ANSWER KEY

Name: ______________________ Date: ______________

A.	B.	C.
Subtract: $\begin{array}{r} 300 \\ -126 \\ \hline 174 \end{array}$	Subtract: $\begin{array}{r} 400 \\ -216 \\ \hline 184 \end{array}$	Subtract: $\begin{array}{r} 400 \\ -258 \\ \hline 142 \end{array}$
D.	**E.**	**F.**
Subtract: $\begin{array}{r} 500 \\ -187 \\ \hline 313 \end{array}$	Subtract: $\begin{array}{r} 300 \\ -214 \\ \hline 86 \end{array}$	Subtract: $\begin{array}{r} 900 \\ -177 \\ \hline 723 \end{array}$
G.	**H.**	**I.**
Subtract: $\begin{array}{r} 300 \\ -176 \\ \hline 124 \end{array}$	Subtract: $\begin{array}{r} 800 \\ -211 \\ \hline 589 \end{array}$	Subtract: $\begin{array}{r} 900 \\ -359 \\ \hline 541 \end{array}$

Level 13 & 14 Mixed Practice Subtracting Multi-Digit Numbers (Regrouping across one or two zeros)
ANSWER KEY

Name: ______________________________ Date: ______________

A.	B.	C.
Subtract: $\begin{array}{r} 401 \\ -383 \\ \hline 18 \end{array}$	Subtract: $\begin{array}{r} 705 \\ -216 \\ \hline 489 \end{array}$	Subtract: $\begin{array}{r} 900 \\ -267 \\ \hline 633 \end{array}$
D.	**E.**	**F.**
Subtract: $\begin{array}{r} 800 \\ -224 \\ \hline 576 \end{array}$	Subtract: $\begin{array}{r} 403 \\ -115 \\ \hline 288 \end{array}$	Subtract: $\begin{array}{r} 302 \\ -184 \\ \hline 118 \end{array}$
G.	**H.**	**I.**
Subtract: $\begin{array}{r} 901 \\ -175 \\ \hline 726 \end{array}$	Subtract: $\begin{array}{r} 700 \\ -223 \\ \hline 477 \end{array}$	Subtract: $\begin{array}{r} 400 \\ -184 \\ \hline 216 \end{array}$

Made in United States
North Haven, CT
10 November 2026

22106134R00096